面向新工科普通高等教育系列教材

# 大数据专业英语教程

张强华  司爱侠  编著

机械工业出版社

本书内容包括大数据基础、大数据的应用、大数据的类型、数据建模、数据采集、云存储、数据库、数据仓库、数据提取、数据转换和加载、大数据分析工具及其主要功能、数据挖掘、数据挖掘算法、大数据编程语言、Spark、数据可视化及其工具、大数据与云计算、大数据安全的挑战与解决方案、大数据与隐私等。

全书共 10 个单元，每一单元包括课文、单词、词组、缩略语、习题、参考译文、阅读材料。附录 A "大数据专业英语词汇特点及翻译技巧"可以帮助读者记忆和翻译专业词汇；附录 B、附录 C、附录 D 既可用于复习和背诵，也可作为小词典长期查阅。本书课文配有音频材料，扫描二维码即可收听。

本书既可作为高等院校本、专科大数据及相关专业的专业英语教材，也可作为相关从业人员自学的参考书，还可作为培训班的培训用书。

本书配有授课电子课件，需要的教师可登录 www.cmpedu.com 免费注册，审核通过后下载，或联系编辑索取（微信：15910938545，电话：010-88379739）。

## 图书在版编目（CIP）数据

大数据专业英语教程 / 张强华，司爱侠编著. —北京：机械工业出版社，2022.6
（2025.1 重印）
面向新工科普通高等教育系列教材
ISBN 978-7-111-70798-1

Ⅰ. ①大…　Ⅱ. ①张…　②司…　Ⅲ. ①数据处理-英语-高等学校-教材

Ⅳ. ①TP274

中国版本图书馆 CIP 数据核字（2022）第 083117 号

机械工业出版社（北京市百万庄大街 22 号　邮政编码 100037）
策划编辑：郝建伟　　责任编辑：郝建伟　解　芳
责任校对：张艳霞　　责任印制：郜　敏

中煤（北京）印务有限公司印刷

2025 年 1 月第 1 版·第 3 次印刷
184mm×260mm·15.25 印张·374 千字
标准书号：ISBN 978-7-111-70798-1
定价：65.00 元

电话服务 网络服务

客服电话：010-88361066　　　机 工 官 网：www.cmpbook.com
　　　　　010-88379833　　　机 工 官 博：weibo.com/cmp1952
　　　　　010-68326294　　　金 书 网：www.golden-book.com
**封底无防伪标均为盗版**　　机工教育服务网：www.cmpedu.com

# 前　言

2015 年，《促进大数据发展行动纲要》发布，十八届五中全会首次提出要实施"国家大数据战略"；2016 年，《政务信息资源共享管理暂行办法》出台；2017 年，《大数据产业发展规划（2016—2020 年）》实施。因此，许多高校陆续开设了大数据专业，培养社会急需的专业人才。而大数据技术的发展极为迅猛，要求从业人员必须掌握许多新技术、新方法，由此也提升了对其专业英语水平的要求。在职场中，具备相关职业技能并精通专业外语的人员往往能赢得竞争优势，成为不可或缺的甚至是引领性的人才。本书旨在帮助读者提高大数据专业英语水平。

本书具有如下特色。

1）本书选材广泛，内容包括大数据基础、大数据的应用、大数据的类型、数据建模、数据采集、云存储、数据库、数据仓库、数据提取、数据转换和加载、大数据分析工具及其主要功能、数据挖掘、数据挖掘算法、大数据编程语言、Spark、数据可视化及其工具、大数据与云计算、大数据安全的挑战与解决方案、大数据与隐私等。

2）体例创新，适合教学。本书内容设计与课堂教学的各个环节紧密切合。本书共 10 个单元，每一单元包含以下部分。课文——选材广泛、风格多样、切合实际的两篇专业文章；单词——给出课文中出现的新词，读者由此可以积累专业词汇；词组——给出课文中的常用词组；缩略语——给出课文中出现的、业内人士必须掌握的缩略语；参考译文——让读者对照理解并提高翻译能力；习题——既有针对课文的练习，也有一些开放性的练习，力求丰富读者的知识储备；阅读材料——这些专业资料可以扩大读者的视野。附录 A "大数据专业英语词汇特点及翻译技巧"可以帮助读者记忆和翻译专业词汇。附录 B、附录 C、附录 D 既可用于复习和背诵，也可作为小词典长期查阅。本书课文配有音频材料，扫描二维码即可收听（音频建议用耳机收听）。

3）教学资源丰富，教学支持完善。本书有配套的教学大纲、教学 PPT、参考答案、参考试卷等资料。需要的教师可登录 www.cmpedu.com 免费注册，审核通过后下载。

本书使用的是英式音标。本书编者有二十余年的专业英语翻译、教学与图书编写经验，出版了多部国家级规划教材，为机械工业出版社金牌作者。读者在使用本书过程中有任何问题，都可以通过电子邮件与编者交流，编者邮箱为 zqh3882355@sina.com，zqh3882355@163.com。

编　者

# 目　录

# Unit 1

## Text A

## Big Data Basics

### 1. What is big data

扫码听课文

According to Gartner, big data is high-volume, velocity and variety information assets that demand cost-effective, innovative forms of information processing for enhanced insight and decision making.

This definition clearly answers the question "What is big data". That is big data refers to complex and large data sets that have to be processed and analyzed to uncover valuable information that can benefit businesses and organizations.

However, there are certain basic tenets of big data that will make it even simpler to answer what big data is.

- It refers to a massive amount of data that keeps on growing exponentially with time.
- It is so voluminous that it cannot be processed or analyzed using conventional data processing techniques.
- It includes data storage, data mining, data analysis, data sharing and data visualization.
- The term is an all-comprehensive one including data, data frameworks, along with the tools and techniques used to process and analyze the data.

### 2. Characteristics of big data

(1) Volume

Big data implies enormous volumes of data. It used to be created by employees. Now that data is generated by machines, networks and human interaction on systems like social media, the volume of data to be analyzed is massive.

(2) Variety

Variety refers to the many sources and types of data both structured and unstructured. We used to store data from sources like spreadsheets and databases. Now data comes in the form of emails, photos, videos, audios, monitoring devices, PDFs, etc. This variety of unstructured data creates problems for storing, mining and analyzing data.

(3) Velocity

Big data velocity deals with the pace at which data flows in from sources like business processes, machines, networks and human interaction with things like social media sites, mobile devices, etc. The flow of data is massive and continuous. This real-time data can help researchers and businesses make valuable decisions that provide strategic competitive advantages and ROI if you are able to handle the velocity.

(4) Veracity

Big data veracity refers to the biases, noise and abnormalities in data. Is the data that is being stored and mined meaningful to the problem being analyzed? In scoping out your big data strategy you need to have your team and partners work to help keep your data clean and processes to keep "dirty data" from accumulating in your systems.

(5) Validity

Like big data veracity is the issue of validity, meaning whether the data is correct and accurate for the intended use. Clearly valid data is key to making the right decisions.

(6) Volatility

Big data volatility refers to how long data is valid and how long it should be stored. In this world of real time data, you need to determine at what point data is no longer relevant to the current analysis.

Big data clearly deals with issues beyond volume, variety and velocity to other concerns like veracity, validity and volatility.

## 3. How does big data work

The main idea behind big data is that the more you know about something, the more you can gain insights and make a decision or find a solution. In most cases this process is completely automated. But to achieve that with the help of analytics tools, machine learning or even artificial intelligence, you need to know how big data works and set up everything correctly.

To handle so much data requires a really stable and well-structured infrastructure. It will quickly process huge volumes and different types of data and this can overload a single server or cluster. That is why you need to have a well-thought out system behind big data.

All the processes should be considered according to the capacity of the system. And this can potentially demand hundreds or thousands of servers for larger companies. As you can imagine, this can start to get pricey. And when you add in all the tools that you will need, it starts to pile up. Therefore, you need to know the three main actions behind big data, so you can plan your budget beforehand and build the best system possible.

(1) Integration

Big data is always collected from many sources and new strategies and technologies to handle enormous loads of information need to be discovered. In some cases, we are talking for petabytes of information flowing into your system, so it will be a challenge to integrate such volume of information in your system. You will have to receive the data, process it and format it in the right form that your business needs and that your customers can understand.

(2) Management

What else might you need for such a large volume of information? You will need a place to store it. Your storage solution can be in the cloud, on-premises, or both. You can also choose in what form your data will be stored, so you can have it available in real-time on-demand. This is why more and more people are choosing a cloud solution for storage because it supports your current compute requirements.

(3) Analysis

After the data has been received and stored, you need to analyze it so you can use it. Explore your data and use it to make any important decisions such as knowing what features are mostly researched from your customers or use it to share research. Do whatever you want and need with it.

## 4. Advantages of big data

(1) Improve business processes

Probably the biggest advantage of big data is that it helps businesses to gain a huge competitive advantage. Apart from being able to understand, as well as target customers better, analyzing big data can result in the improvement and optimization of certain facets of business operations. For instance, by mining big data, retailers can not only explore patterns in consumption and production but can also promote better inventory management, improve the supply chain, optimize distribution channels, among others.

(2) Detect frauds

This advantage of big data comes from the implementation of machine learning technologies. It helps banks and other financial institutions to detect frauds like fraudulent purchases with credit cards often before even the cardholder gets to know about it.

(3) Improve customer services

One of the most common goals among big data analytics programs is improving customer services. Today's businesses capture a huge amount of information from different sources like Customer Relationship Management (CRM) systems, social media together with other points of customer contact. By analyzing this massive amount of information they get to know about the tastes and preferences of a user. And with the help of the big data technology, they are able to create experiences which are more responsive, personal and accurate than ever before.

## 5. Disadvantages of big data

Despite the advantages of big data, it comes with some serious challenges that make its implementation difficult or risky.

(1) Privacy and security concerns

Probably the biggest disadvantage of big data is that it can make businesses a target for cyber attackers. Even giant businesses have experienced instances of massive data breaches. However, with the implementation of GDPR, businesses are increasingly trying to invest in processes, protocols and infrastructure to be able to maintain big data.

(2) Need for technical expertise

Working with big data needs a great deal of technical skills and that's one of the key reasons why big data experts and data scientists belong to the highly paid and highly coveted group in the IT field. Training existing staff or hiring experts to handle big data can easily increase the cost of a business considerably.

## 6. Where is big data headed in the future

Big data has changed the rules game in many fields and will undoubtedly continue to grow. Once everything around us starts using the Internet (Internet of Things), the possibilities of using big data will be enormous. The amount of data available to us is only going to increase, and analytics technology will become more advanced.

All the tools used for big data are going to evolve as well. The infrastructure requirements are going to change. Maybe in the future we will be able to store all the data we need on only one machine. This could potentially make everything cheaper and easier to work with.

## ✎ New Words

| volume | ['vɒljuːm] | n.容量，大量 |
|---|---|---|
| velocity | [vəˈlɒsəti] | n.速率，速度；高速，快速 |
| variety | [vəˈraɪəti] | n.多样化；种类 |
| innovative | [ˈɪnəveɪtɪv] | adj.革新的，创新的 |
| enhance | [ɪnˈhɑːns] | vt.提高，增加；加强 |
| uncover | [ʌnˈkʌvə] | vi.发现，揭示 |
| valuable | [ˈvæljuəbl] | adj.有价值的 |
| tenet | [ˈtenɪt] | n.原则；信条 |
| massive | [ˈmæsɪv] | adj.大量的；大规模的 |
| exponentially | [ˌekspəˈnenʃəli] | adv.以指数方式 |
| voluminous | [vəˈluːmɪnəs] | adj.大的；多产的 |
| process | [ˈprəʊses] | vt.加工；处理<br>n. 步骤，程序；过程 |
| analyze | [ˈænəlaɪz] | vt.分析；分解 |
| conventional | [kənˈvenʃnl] | adj.传统的；依照惯例的，平常的 |
| storage | [ˈstɔːrɪdʒ] | n.存储 |
| framework | [ˈfreɪmwɜːk] | n.架构；框架；（体系的）结构 |
| characteristic | [ˌkærəktəˈrɪstɪk] | n.特性，特征，特色<br>adj.特有的；独特的 |
| spreadsheet | [ˈspredʃiːt] | n.电子制表软件 |
| email | [ˈiːmeɪl] | n.电子邮件<br>vt.给……发电子邮件 |
| continuous | [kənˈtɪnjuəs] | adj.连续的，不断的 |
| handle | [ˈhændl] | vi.处理；操作，操控<br>n.把手；方法；提手 |
| veracity | [vəˈræsəti] | n.真实性 |
| bias | [ˈbaɪəs] | n.偏见；偏爱；倾向 |
| noise | [nɔɪz] | n.噪声 |

| | | |
|---|---|---|
| abnormality | [ˌæbnɔ:'mæləti] | n.异常；变态；畸形 |
| accumulate | [ə'kju:mjəleɪt] | v.堆积，积累 |
| validity | [və'lɪdəti] | n.有效性，合法性 |
| volatility | [ˌvɒlə'tɪlɪti] | n.易变性 |
| valid | ['vælɪd] | adj.有效的 |
| automate | ['ɔ:təmeɪt] | v.（使）自动化 |
| stable | ['steɪbl] | adj.稳定的；持久的 |
| overload | [ˌəʊvə'ləʊd] | vt.使负担太重；使超载；超过负荷 |
| budget | ['bʌdʒɪt] | n.预算 |
| | | v.把……编入预算 |
| integration | [ˌɪntɪ'greɪʃn] | n.整合；一体化 |
| strategy | ['strætədʒi] | n.策略，战略 |
| available | [ə'veɪləbl] | adj.可获得的；有空的；能找到的 |
| explore | [ɪk'splɔ:] | vt.探索，探究 |
| improvement | [ɪm'pru:vmənt] | n.改进之处，改善 |
| consumption | [kən'sʌmpʃn] | n.消费 |
| production | [prə'dʌkʃn] | n.生产，制作；产品；产量 |
| promote | [prə'məʊt] | vt.促进，推进 |
| fraud | [frɔ:d] | n.欺诈；骗子 |
| fraudulent | ['frɔ:djələnt] | adj.欺骗的，不诚实的 |
| cardholder | ['kɑ:dhəʊldə] | n.持有信用卡的人 |
| program | ['prəʊgræm] | n.程序；计划，安排 |
| | | v.给……编写程序；为……制订计划 |
| responsive | [rɪ'spɒnsɪv] | adj.响应的 |
| accurate | ['ækjʊrət] | adj.精确的，准确的 |
| serious | ['sɪərɪəs] | adj.严重的；重要的；危急的 |
| breach | [bri:tʃ] | vt.攻破；破坏 |
| | | n.破坏；破裂 |
| protocol | ['prəʊtəkɒl] | n.协议 |
| expertise | [ˌekspɜ:'ti:z] | n.专门知识或技能；专家的意见 |
| proficiency | [prə'fɪʃnsi] | n.熟练，精通，娴熟 |
| expert | ['ekspɜ:t] | n.专家，能手；权威；行家 |
| | | adj.专业的 |
| undoubtedly | [ʌn'daʊtɪdli] | adv.毋庸置疑地，的确地；显然，必定 |
| enormous | [ɪ'nɔ:məs] | adj.巨大的，庞大的 |
| possibility | [ˌpɒsə'bɪləti] | n.可能性 |
| potentially | [pə'tenʃəli] | adv.潜在地，可能地 |

## ✎ Phrases

| | |
|---|---|
| big data | 大数据 |
| information processing | 信息处理 |
| data processing technique | 数据处理技术 |
| data storage | 数据存储 |
| data mining | 数据挖掘 |
| data analysis | 数据分析 |
| data sharing | 数据共享 |
| data visualization | 数据可视化 |
| social media | 社交媒体 |
| monitoring device | 监控设备 |
| data flow | 数据流 |
| mobile device | 移动设备 |
| dirty data | 脏数据，废数据 |
| real time | 实时 |
| well-thought out | 深思熟虑 |
| pile up | 堆积，积聚，成堆 |
| competitive advantage | 竞争优势 |
| inventory management | 库存管理 |
| supply chain | 供应链 |
| distribution channel | 分销渠道 |
| credit card | 信用卡 |
| Internet of Things | 物联网 |

## ✎ Abbreviations

| | |
|---|---|
| ROI (Return on Investment) | 投资利润，投资回报率 |
| CRM (Customer Relationship Management) | 客户关系管理 |
| GDPR (General Data Protection Regulation) | 通用数据保护条例 |
| IT(Information Technology) | 信息技术 |

## Reference Translation

# 大数据基础

### 1. 什么是大数据

根据 Gartner 的说法，大数据是大容量、快速和多样的信息资产，它们需要经济、高效、创新的信息处理形式，以增强洞察力和决策能力。

这个定义清楚地回答了"什么是大数据"这一问题。大数据是指必须处理和分析复杂且巨大的数据集，以发现可以使企业和组织受益的有价值信息。

但是，大数据有一些基本原则，这使得回答什么是大数据变得更加简单。

- 它是指大量数据，其随着时间呈指数增长。
- 它是如此庞大，以至于无法使用常规数据处理技术进行处理或分析。
- 它包括数据存储、数据挖掘、数据分析、数据共享和数据可视化。
- 该术语是一个全面的术语，包括数据、数据框架以及用于处理和分析数据的工具与技术。

## 2. 大数据的特征

（1）数据量大

大数据意味着海量数据。它曾经是员工创建的数据。现在，因为数据是由机器、网络和诸如社交媒体之类的系统上的人员交互生成的，因此要分析的数据量很大。

（2）多样性

多样性指结构化和非结构化数据有许多来源与类型。人们曾经存储来自电子表格和数据库等来源的数据。现在，数据以电子邮件、照片、视频、音频、监视设备、PDF 等形式出现。各种各样的非结构化数据给存储、挖掘和分析数据带来了麻烦。

（3）高速性

大数据的高速性是指处理数据从业务流程、机器、网络以及人类与社交媒体站点交互、移动设备等流入的速度。数据流很大而且是连续的。这些实时数据可以帮助研究人员和企业做出有价值的决策，如果用户能够掌握数据流入的速度，则可以提供战略竞争优势和投资回报率。

（4）真实性

大数据的真实性是指数据中的偏差、噪声和异常。存储和挖掘的数据是否对正在分析的问题有意义？在制定大数据策略时，需要团队和合作伙伴的共同努力，以保持数据干净并防止"脏数据"在系统中累积。

（5）有效性

就像大数据的真实性一样，有效性问题也意味着数据对于预期用途而言是正确且准确的。明确有效的数据是做出正确决策的关键。

（6）易变性

大数据的易变性是指数据有效期为多长时间以及数据应存储多长时间。在实时数据的世界中，需要确定数据在什么时候不再与当前分析相关。

显然，大数据涉及的问题不仅限于数量、多样性和高速性，还涉及其他问题，如真实性、有效性和易变性。

## 3. 大数据如何工作

大数据背后的主要思想是：对某一事情了解得越多，就越能获得洞察力，更能做出决定或找到解决方案。在大多数情况下，此过程是完全自动化的。但要在分析工具、机器学习甚至人工智能的帮助下实现这一目标，需要了解大数据的工作原理并正确设置一切。

处理大量数据需要一个真正稳定且结构良好的基础架构。它将快速处理海量和不同类型的数据，这可能会使单个服务器或集群过载。这就是为什么需要在大数据背后拥有一个经过深思熟虑的系统。

应该根据系统容量考虑所有过程。对于大型公司来说，这可能需要成百上千台服务器。正如想象的那样，这可能要花费大量的资金。当添加所需的所有工具时，费用就多了。因此，需要了解大数据三个主要操作，以便可以事先计划预算，并构建最佳的系统。

（1）整合

大数据总是从许多来源收集的，因此需要找到处理巨量信息的新策略和技术。在某些情况下，人们正在谈论流入系统的 PB 级信息，因此将如此大量的信息集成到系统中将是一个挑战。必须按照业务需要和客户可以理解的正确格式来接收、处理和格式化数据。

（2）管理

如此大量的信息还需要什么？需要一个存放它的地方。信息可以存储在云中、本地或同时在两者中。还可以选择以什么形式存储数据，以便可以按需实时提供数据。这就是越来越多的人选择云存储解决方案的原因，因为它支持当前的计算要求。

（3）分析

在接收并存储了数据之后，还需要对其进行分析以便使用。探索数据并使用它来做出任何重要的决定，例如了解客户主要研究了哪些功能或使用它来共享研究。用户可以使用它来实现想做的和需要的一切。

## 4. 大数据的优势

（1）改善业务流程

大数据的最大优势可能是它可以帮助企业获得巨大的竞争优势。除了能够更好地了解和定位目标客户之外，分析大数据还可以改善和优化业务运营的某些方面，如通过挖掘大数据，零售商不仅可以探索消费和生产方式，还可以促进库存管理、改善供应链、优化分销渠道等。

（2）检测欺诈

大数据的这个优势来自机器学习技术的实施。它可以帮助银行和其他金融机构经常在甚至连持卡人都不知道的情况下，检测出像用信用卡欺诈购买等欺诈行为。

（3）改善客户服务

大数据分析程序中最常见的目标之一是改善客户服务。当今的企业从不同的来源（如客户关系管理（CRM）系统、社交媒体以及其他客户联系点）捕获大量信息。通过分析大量信息，他们可以了解用户的口味和喜好。在大数据技术的帮助下，他们能够创造比以往任何时候都更加响应灵敏、个性化和准确的客户体验。

## 5. 大数据的缺点

尽管大数据有很多优势，但它仍然面临着一些严峻的挑战，这些挑战使其实施变得困难或存在风险。

（1）隐私和安全问题

大数据最大的缺点可能是它会使企业成为网络攻击者的目标，甚至大型企业也经历过大规模数据泄露的事件。但是，随着 GDPR 的实施，企业越来越多地尝试在流程、协议和基础架构上进行投资，以便维护大数据。

（2）对技术专业知识的需求

处理大数据需要大量的技术能力，这是大数据专家和数据科学家成为 IT 领域收入丰厚且令人羡慕的团队的主要原因之一。培训员工或雇用专家来处理大数据可能会轻易地增加企业成本。

**6. 大数据未来的发展方向在哪**

大数据已经在许多领域改变了游戏规则，而且无疑将继续增长。一旦我们周围的一切都开始使用互联网（物联网），那么使用大数据的可能性将是巨大的。我们可得到的数据量只会增加，而分析技术将会变得更加先进。

所有用于大数据的工具也将不断发展，基础架构要求也将发生变化。也许未来我们能够将所需的所有数据存储在一台机器上。这可能会使一切变得更便宜且更容易使用。

# Text B

## Applications of Big Data

扫码听课文

In today's world, there are a lot of data. Big companies utilize those data for their business growth. By analyzing those data, they can make wise decisions in various cases.

**1. Tracking customer's spending habits and shopping behaviors**

In big retails stores like Amazon, Walmart, Big Bazar etc. the management teams have to keep data of customer's spending habits (in which product the customer buys, in which band he wishes to spend, how frequently he spends), shopping behavior, customer's most favourite product and which product is being searched/sold most. Based on that data, they can fix the production/collection rate of that product.

Banking sector uses their customer's spending behavior-related data so that they can provide the offer to a particular customer to buy his particular liked product by using bank's credit or debit card with discount or cashback. In this way, they can send the right offer to the right person at the right time.

**2. Recommendation**

By tracking customer's spending habits, shopping behaviors, big retails store provide a recommendation to the customer. E-commerce sites like Amazon, Walmart, Flipkart carry out product recommendation. They track what products a customer is searching, and based on that data they recommend that type of products to that customer.

Suppose a customer searched bed cover on Amazon. Amazon got the data that customer may be interested in buying bed cover. Later, when that customer goes to any Google page, he will see advertisements of various bed covers. Thus, advertisements of the right products can be sent to the right customers.

YouTube also recommends videos based on what video a user previously liked, or watched. Based on the content of a video the user is watching, relevant advertisement is shown during video running. For example, when someone is watching a tutorial video of big data, advertisements of other big data courses will be shown during that video.

### 3. Smart traffic system

Data about the condition of the traffic of different roads, data collected through camera kept beside the roads, at entry and exit point of the city, and GPS device placed in the vehicle (Ola, Uber cab, etc.). All such data are analyzed and people are recommended a way that is jam-free or less jammed, taking less time. Smart traffic system can be built in the city by big data analysis. One more profit is that fuel consumption.

### 4. Secure air traffic system

Sensors are present at various places of the plane (like propeller etc). These sensors can capture data like the speed of flight, moisture, temperature and other environmental condition. Based on such data analysis, an environmental parameter within flight are set up and varied. By analyzing flight's machine-generated data, it can be estimated how long the machine can operate flawlessly before it is to be replaced/repaired.

### 5. Auto driving car

Big data analysis helps drive a car without human interpretation. In the various spots of car camera, a sensor is placed. It gathers data like the size of the surrounding cars, obstacle, distance from those, etc. These data are being analyzed, then various calculations like how many angles to rotate, what should be the speed, when to stop, etc. are carried out. These calculations help to take action automatically.

### 6. Virtual personal assistant tool

Big data analysis helps virtual personal assistant tool (like Siri in Apple Device, Cortana in Windows, Google Assistant in Android) to provide the answers to the various question asked by users. This tool tracks the location of the user, their local time, season, other data related to question asked, etc. By analyzing all such data, it provides an answer.

As an example, suppose one user asks "Do I need to take an umbrella", the tool collects data like location of the user, season and weather condition at that location, then it analyzes these data to conclude if there is a chance of raining, and finally provides the answer.

### 7. IoT

Manufacturing companies install IoT sensor into machines to collect operational data. By analyzing such data, the company can predict how long a machine will work without any problem before it requires repairing so that company can take actions before the machine faces a lot of issues or gets totally down, thus saving the cost to replace the whole machine.

In the healthcare field, big data is providing a significant contribution. IoT sensors placed near the patient can constantly keep track of various health conditions of the patient like heart bit rate, blood presser, etc. Whenever any parameter crosses the safe limit, an alarm is sent to a doctor, so they can take steps remotely as soon as possible.

### 8. Education sector

Online educational course conducting organization utilize big data to search candidates interested in that course. If someone searches for YouTube tutorial video on a subject, online or offline course provider organizations on that subject will send online advertising to that person about their course.

## 9. Energy sector

Smart electric meter reads consumed power every 15 minutes and sends this read data to the server. Then the server analyzes the data and estimates what is the time in a day when the power load is less throughout the city. The system suggests manufacturing units or housekeepers the time they should drive their heavy machine is the night time, when power load is less and they can enjoy less electricity bill.

## 10. Media and entertainment sector

Media and entertainment service providing companies like Netflix, Amazon Prime, Spotify do analysis on data collected from their users. They collect and analyze data like what type of video users are watching most, what music they are listening most, how long users are spending on site, etc. to set the next business strategy.

## New Words

| | | |
|---|---|---|
| habit | ['hæbɪt] | n.习惯，习性 |
| favourite | ['feɪvərɪt] | adj.特别受喜爱的<br>n.特别喜爱的人（或物） |
| sector | ['sektə] | n.部门；领域 |
| particular | [pə'tɪkjələ] | adj.特别的；详细的；独有的<br>n.细节；详情 |
| cashback | ['kæʃbæk] | n.现金返还 |
| recommendation | [ˌrekəmen'deɪʃn] | n.推荐 |
| track | [træk] | vt.跟踪；监看，监测<br>n.小路；踪迹 |
| advertisement | [əd'vɜːtɪsmənt] | n.广告，公告 |
| camera | ['kæmərə] | n.摄像头；照相机；摄影机 |
| vehicle | ['viːəkl] | n.车辆，交通工具 |
| fuel | ['fjuːəl] | n.燃料 |
| propeller | [prə'pelə] | n.螺旋桨，推进器 |
| moisture | ['mɔɪstʃə] | n.水分；湿气 |
| environmental | [ɪnˌvaɪrən'mentl] | adj.自然环境的 |
| parameter | [pə'ræmɪtə] | n.参数 |
| replace | [rɪ'pleɪs] | vt.替换；代替 |
| repair | [rɪ'peə] | vt.修理；纠正；恢复 |
| interpretation | [ɪnˌtɜːprɪ'teɪʃn] | n.解释，说明 |
| surrounding | [sə'raʊndɪŋ] | adj.周围的，附近的<br>n.周围环境 |
| obstacle | ['ɒbstəkl] | n.障碍（物） |
| rotate | [rəʊ'teɪt] | v.（使某物）旋转；使转动 |

| calculation | [ˌkælkjʊ'leɪʃn] | *n.*计算 |
|---|---|---|
| contribution | [ˌkɒntrɪ'bjuːʃn] | *n.*贡献 |
| treatment | ['triːtmənt] | *n.*治疗；处理 |
| symptom | ['sɪmptəm] | *n.*症状；征兆 |
| disease | [dɪ'ziːz] | *n.*疾病；弊端 |
| alarm | [ə'lɑːm] | *n.*警报；闹钟 |
| | | *vt.*警告 |
| housekeeper | ['haʊskiːpə] | *n.*管家；主妇 |

## ✍ Phrases

| a lot of | 许多的 |
|---|---|
| in various cases | 在各种情况下 |
| spending habit | 消费习惯 |
| debit card | 借记卡 |
| e-commerce site | 电子商务网站 |
| carry out | 执行；进行 |
| be interested in | 对……感兴趣 |
| smart traffic system | 智能交通系统 |
| virtual personal assistant | 虚拟个人助理 |
| heart bit rate | 心率 |
| smart electric meter | 智能电表 |

## Reference Translation

### 大数据的应用

在当今世界上有很多数据，大公司利用这些数据来促进业务增长。通过分析这些数据，他们可以在各种情况下做出明智的决策。

**1. 跟踪客户的消费习惯和购物行为**

在亚马逊、沃尔玛和 Big Bazar 等大型零售商店中，管理团队必须保存客户消费习惯（客户购买哪种产品、他希望消费的价格范围及消费频率）、购物行为、最喜欢的产品以及最频繁搜索/销售的产品等数据。根据该数据，他们可以修订该产品的生产率/账款回收率。

银行部门使用与其客户的消费行为相关的数据，能够向特定客户提供他喜欢的产品报价，客户可以使用带有折扣或现金返还的银行信用卡或借记卡购买。这样，他们可以在合适的时间向适当的人发送正确的报价。

**2. 推荐**

通过跟踪客户的消费习惯及购物行为，大型零售商店会向客户提供推荐。像亚马逊、沃尔玛和 Flipkart 等电子商务网站会推荐产品。他们跟踪客户正在搜索的产品，并根据这些数据向该客户推荐该类型的产品。

假设有一位顾客在亚马逊上搜索了床罩，亚马逊得到该客户可能有兴趣购买床罩的数据。之后，在该客户访问任何 Google 页面时，将看到各种床罩的广告。这样，适当的产品广告就发送给了合适的顾客。

YouTube 还根据用户之前喜欢或观看过的视频来推荐视频。根据用户正在观看的视频内容，在视频播放过程中会显示相关的广告。例如，当某人观看大数据课程的视频时，在该视频播放期间将显示其他大数据课程的广告。

## 3. 智能交通系统

有关不同道路交通状况的数据、通过道路旁的摄像头收集的数据、城市出入口处的数据，以及放置在车辆（Ola、Uber 出租车等）中的 GPS 设备的数据。如果对这些数据进行分析，就能给人们提供无阻塞或较少阻塞的交通方式，从而节省时间。通过大数据分析，可以在城市中建立智能交通系统。另一个好处是可以减少燃油消耗。

## 4. 安全的空中交通系统

传感器存在于飞机的各个位置（如螺旋桨等）。这些传感器可以捕获数据，如飞行速度、湿度、温度及其他环境条件。基于对此类数据的分析，可以设置并改变飞行中的环境参数。通过分析飞行器生成的数据，可以估算机器在更换/维修之前可以无误运行多长时间。

## 5. 自动驾驶汽车

大数据分析有助于驾驶汽车而无需人工判读。在汽车摄像头的各个位置，都放置了一个传感器，它收集诸如周围汽车的大小、障碍物、与车辆或障碍物之间的距离等数据。通过对这些数据进行分析，就能进行各种计算（如旋转多少角度、速度应该是多少、何时停止等）。这些计算有助于自动采取措施。

## 6. 虚拟个人助理工具

大数据分析可以帮助虚拟个人助理工具（如 Apple Device 中的 Siri、Windows 中的 Cortana、Android 中的 Google Assistant）为用户提出的各种问题提供答案。这个虚拟的个人助理工具可以跟踪用户的位置、他们的本地时间、季节、与所问问题相关的其他数据等。通过分析所有此类数据，它可以提供答案。

例如，假设一个用户问"我需要带雨伞吗？"，该工具会收集用户所在位置、季节和该位置的天气状况等数据，然后分析这些数据以得出是否会下雨，并最后提供答案。

## 7. 物联网

制造公司将物联网传感器安装到机器中以收集运行数据。通过分析这些数据，公司可以预测机器在需要维修之前可以正常工作多长时间，使公司可以在机器面临很多问题或完全瘫痪之前采取行动。这样可以节省更换整个机器的成本。

在医疗保健领域，大数据正在发挥重要作用。放置在患者附近的物联网传感器可以持续跟踪患者的各种健康状况（如心率、血压等）。只要任何参数超过安全极限，就会向医生发送警报，这样他们可以尽快远程采取措施。

## 8. 教育领域

在线教育课程组织机构利用大数据来搜索对该课程感兴趣的人员。如果有人搜索有关某个主题的 YouTube 教程视频，则有关该主题的在线或离线课程提供商将在线向其发送有关该课程的广告。

## 9. 能源领域

智能电表每隔 15 分钟读取一次用电量，并将读取的数据发送到服务器，服务器对数据进行分析，然后可以估算出一天中整个城市用电负荷较少的时段。该系统建议制造厂或管家在夜间使用重型机器，这时用电负荷较小，他们要交的电费也少。

## 10. 媒体和娱乐领域

Netflix、Amazon Prime、Spotify 等媒体和娱乐服务提供公司会对从其用户那里收集的数据进行分析。他们收集并分析用户观看的视频类型、最常听的音乐、用户在网站上花费的时间等数据，以设置下一个业务策略。

# Exercises

[Ex. 1] 根据 Text A 回答以下问题。

1. What is big data according to Gartner?

2. What are the characteristics of big data?

3. What does variety refer to?

4. What does big data velocity do?

5. What is the main idea behind big data?

6. What are the three main actions you need to know behind big data?

7. What is probably the biggest advantage of big data?

8. What is one of the most common goals among big data analytics programs?

9. What is probably the biggest disadvantage of big data?

10. What is one of the key reasons why big data experts and data scientists belong to the highly paid and highly coveted group in the IT field?

[Ex. 2] 根据 Text B 回答以下问题。

1. What do the management teams have to in big retails stores like Amazon, Walmart, Big Bazar etc.?

2. What is the purpose of banking sector to use their customer's spending behavior-related data?

3. How do big retail stores provide a recommendation to the customer?

4. How can smart traffic system be built in the city? What is one more profit?

5. What can be estimated by analyzing flight's machine-generated data?

6. What does big data analysis help virtual personal assistant tool (like Siri in Apple Device, Cortana in Windows, Google Assistant in Android) to do?

7. What can IoT sensors placed near the patient do constantly?

8. If someone searches for YouTube tutorial video on a subject, what will online or offline course provider organizations on that subject do?

9. When does the system suggest that manufacturing units or housekeepers should drive their heavy machine? Why?

10. What kind of data do media and entertainment service providing companies collect and analyze to set the next business strategy?

**[Ex. 3]** 词汇英译中

| | |
|---|---|
| 1. automate | 1. _____ |
| 2. enhance | 2. _____ |
| 3. framework | 3. _____ |
| 4. data mining | 4. _____ |
| 5. parameter | 5. _____ |
| 6. data storage | 6. _____ |
| 7. calculation | 7. _____ |
| 8. data analysis | 8. _____ |
| 9. particular | 9. _____ |
| 10. data visualization | 10. _____ |

**[Ex. 4]** 词汇中译英

| | |
|---|---|
| 1. 脏数据，废数据 | 1. _____ |
| 2. 多样化；种类 | 2. _____ |
| 3. 社交媒体 | 3. _____ |
| 4. 易变性 | 4. _____ |

5. 真实性                       5. _____

6. 虚拟个人助理         6. _____

7. 移动设备               7. _____

8. 容量，大量           8. _____

9. 存储                    9. _____

10. 稳定的；持久的     10. _____

## [Ex. 5] 短文翻译

### Big Data Glossary

Batch processing: Batch processing is a computing strategy that involves processing data in large sets. This is typically ideal for non-time sensitive work that operates on very large sets of data. The process is started and at a later time, the results are returned by the system.

Cluster computing: Cluster computing is the practice of pooling the resources of multiple machines and managing their collective capabilities to complete tasks. Computer clusters require a cluster management layer which handles communication between the individual nodes and coordinates work assignment.

Data lake: Data lake is a term for a large repository of collected data in a relatively raw state. This is frequently used to refer to the data collected in a big data system which might be unstructured and frequently changing. This differs in spirit to data warehouses.

Data mining: Data mining is a broad term for the practice of trying to find patterns in large sets of data. It is the process of trying to refine a mass of data into a more understandable and cohesive set of information.

Data warehouse: Data warehouses are large, ordered repositories of data that can be used for analysis and reporting. In contrast to a data lake, a data warehouse is composed of data that has been cleaned, integrated with other sources, and is generally well-ordered. Data warehouses are often related to big data, but typically are components of more conventional systems.

In-memory computing: In-memory computing is a strategy that involves moving the working datasets entirely within a cluster's collective memory. Intermediate calculations are not written to disk and are instead held in memory. This gives in-memory computing systems like Apache Spark a huge advantage in speed over I/O bound systems like Hadoop's MapReduce.

Stream processing: Stream processing is the practice of computing over individual data items as they move through a system. This allows for real-time analysis of the data being fed to the system and is useful for time-sensitive operations using high velocity metrics.

# Reading Material

## The Future of Big Data

In 2020, every person in the world will be creating 7 MBs of data every second. We have already created more data in past couple of years than in the entire history of human kind. Big data has taken the world by storm and there are no signs of slowing down. You might be thinking, "Where would big data industry go from here?"

## 1. Machine learning will be the next big thing in big data

One of the hottest technology trends today is machine learning and it will play a big part in the future of big data as well. According to Ovum, machine learning will be at the forefront[1] of the big data revolution. It will help businesses in preparing data and conduct predictive analysis so that businesses can overcome future challenges easily.

## 2. Privacy will be the biggest challenge

Whether it is the Internet of Things[2] or big data, the biggest challenge for emerging technologies has been security and privacy of data. The volume of data we are creating right now and the volume of data that will be created in the future will make privacy even more important as stakes[3] will be much higher. Data security and privacy concerns will be the biggest hurdle[4] for big data industry.

## 3. Chief data officer: a new position will emerge

You might be familiar with Chief Executive Officer (CEO), Chief Marketing Officer[5] (CMO) and Chief Information Officer (CIO), but have you ever heard about Chief Data Officer (CDO)? If your answer is no, do not worry because you will soon come to know about it. According to Forrester, we will see the emergence of CDO as the new position and businesses will appoint CDO. Although the appointment of CDO solely depends on the type of business and its data needs but the wider adoption of big data technologies across enterprises, hiring a CDO will become the norm[6].

## 4. Data scientists will be in high demand[7]

If you are still not quite sure about which career path to choose then, there is no better time to start your career in data sciences. As the volume of data grows and big data grows bigger, demand for data scientists, analysts and data management experts will shoot up. The gap between the demand for data professionals and the availability will widen. This will help data scientists and analysts draw higher salaries.

## 5. Businesses will buy algorithms instead of software

More and more businesses will look to purchase algorithm instead of creating their own. After buying an algorithm, businesses can add their own data to it. It provides businesses with more customization options as compared to when they are buying software. You cannot tweak[8] software according to your needs. In fact, it is the other way around. Your business will have to adjust according

---

1　forefront ['fɔ:frʌnt] n.最前部；前列；第一线
2　Internet of Things: 物联网
3　stake [steɪk] n.股份；利害关系
4　hurdle ['hɜ:dl] n.障碍，困难
5　Chief Marketing Officer: 首席营销官
6　norm [nɔ:m] n.常态；标准；规范
7　demand [dɪ'mɑ:nd] n.要求；需求
8　tweak [twi:k] vt.稍稍调整（机器、系统等）

to the software processes, but all this will end soon with algorithms selling services taking center stage.

## 6. Investments in big data technologies will skyrocket

Although the business investments in big data might vary from industry to industry, the increase in big data spending will remain consistent overall. Manufacturing industry will spend the most on big data technology while health care, banking and resource industries will be the fastest to adopt.

## 7. More developers will join the big data revolution

According to statistics, there are six million developers currently working with big data and using advanced analytics. This makes up more than 33% of developers in the world. What's even more amazing is that big data is just getting starting so we will see a surge[9] in a number of developer developing applications for big data in years to come.

## 8. Big data will help you break productivity records

None of your future investments will deliver a higher return on your investment than if you invest in big data, especially when it comes to boosting your business productivity. To give you a better idea, let us put numbers into perspective[10]. According to IDC, organizations that invest in this technology and attain[11] capabilities to analyze large amounts of data quickly and extract actionable information can get an extra $430 billion in terms of productivity benefits over their competitors.

---

9　surge [sɜːdʒ] *n.&v.*汹涌；激增

10　perspective [pəˈspektɪv] *n.*观点，看法；远景

11　attain [əˈteɪn] *v.*获得，得到

# Unit 2

## Text A

## Structured Data, Unstructured Data and Semi-structured Data

### 1. Structured data

Structured data is the data which conforms to a data model, has a well-defined structure, follows a consistent order and can be easily accessed and used by a person or a computer program. It is usually stored in well-defined schemas such as databases. It is generally tabular with column and rows that clearly define its attributes. SQL is often used to manage structured data stored in databases.

扫码听课文

(1) Characteristics of structured data

- Data conforms to a data model and has easily identifiable structure.
- Data is stored in the form of rows and columns.
- Data is well-organized, so the definition, format and meaning of data are explicitly known.
- Data resides in fixed fields within a record or file.
- Similar entities are grouped together to form relations or classes.
- Entities in the same group have same attributes.
- It is easy to access and query, so data can be easily used by other programs.

(2) Sources of structured data

- SQL databases.
- Spreadsheets such as Excel.
- OLTP systems.
- Online forms.
- Sensors such as GPS or RFID tags.
- Network and Web server logs.
- Medical devices.

(3) Advantages of structured data

- Structured data have a well-defined structure that helps in easy storage and access of data.
- Data can be indexed based on text strings as well as attributes. This makes search operation hassle-free.
- Data mining is easy. Knowledge can be easily extracted from data.
- Operations such as updating and deleting are easy due to well structured form of data.

- Business intelligence operations such as data warehousing can be easily undertaken.
- It is easily scalable in case there is an increment of data.
- Ensuring security to data is easy.

Note: Structured data accounts for only about 20% of data but its high degree of organization and performance make it foundation of big data.

## 2. Unstructured data

Unstructured data is the data which does not conforms to a data model and has no easily identifiable structure such that it can not be used by a computer program easily. Unstructured data is not organized in a pre-defined manner nor has a pre-defined data model, thus it is not a good fit for a mainstream relational database.

(1) Characteristics of unstructured data

- Data neither conforms to a data model nor has any structure.
- Data can not be stored in the form of rows and columns.
- Data does not follows any semantic or rules.
- Data lacks any particular format or sequence.
- Data has no easily identifiable structure.
- Due to lack of identifiable structure, it can not used by computer programs easily.

(2) Sources of unstructured data

- Web pages.
- Images (JPEG, GIF, PNG, etc.).
- Videos.
- Memos.
- Reports.
- Word documents and PowerPoint presentations.
- Surveys.

(3) Advantages of unstructured data

- It supports the data which lacks a proper format or sequence.
- The data is not constrained by a fixed schema.
- It is very flexible due to absence of schema.
- Data is portable.
- It is very scalable.
- It can easily deal with the heterogeneity of data sources.
- It has a variety of business intelligence and analytics applications.

(4) Disadvantages of unstructured data

- It is difficult to store and manage unstructured data due to lack of schema and structure.
- Indexing the data is difficult and error prone due to unclear structure and not having pre-defined attributes, due to which search results are not very accurate.
- Ensuring security to data is a difficult task.

(5) Problems faced in storing unstructured data

- It requires a lot of storage space to store unstructured data.
- It is difficult to store videos, images, audios, etc.
- Due to unclear structure, operations like update, delete and search are very difficult.
- Storage cost is high as compared to structured data.
- Indexing the unstructured data is difficult.

(6) Possible solutions for storing unstructured data

- Unstructured data can be converted to easily manageable formats.
- Use Content Addressable Storage (CAS) system to store unstructured data.
- It stores data based on their metadata and a unique name is assigned to every object stored in it. The object is retrieved based on content not its location.
- Unstructured data can be stored in XML format.
- Unstructured data can be stored in RDBMS which supports BLOBs.

(7) Extracting information from unstructured data

Unstructured data does not have any structure. So it can not be easily interpreted by conventional algorithms. It is also difficult to tag and index unstructured data. So extracting information from it is tough job. Here are some possible solutions.

- Taxonomies or classification of data helps in organizing data in hierarchical structure, which will make search process easy.
- Data can be stored in virtual repository and be automatically tagged.
- Use of application platforms like XOLAP. XOLAP helps in extracting information from e-mails and XML based documents.
- Various data mining tools are used.

## 3. Semi-structured data

Semi-structured data is the data which does not conforms to a data model but has some structure. It lacks a fixed or rigid schema. It is the data that does not reside in a rational database but that have some organizational properties that make it easier to analyse. With some process, we can store them in the relational database.

(1) Characteristics of semi-structured data

- Data does not conforms to a data model but has some structure.
- Data can not be stored in the form of rows and columns.
- Semi-structured data contains tags and elements (metadata), which are used to group data and describe how the data is stored.
- Similar entities are grouped together and organized in a hierarchy.
- Entities in the same group may or may not have the same attributes.
- It does not contain sufficient metadata, which makes automation and management of data difficult.
- Size and type of the same attributes in a group may differ.
- Due to lack of a well-defined structure, it can not used by computer programs easily.

(2) Sources of semi-structured data

- E-mails.
- XML and other markup languages.
- Binary executables.
- TCP/IP packets.
- Zipped files.
- Integration of data from different sources.
- Web pages.

(3) Advantages of semi-structured data

- The data is not constrained by a fixed schema.
- Flexible, Schema can be easily changed.
- Data is portable.
- It is possible to view structured data as semi-structured data.
- It supports users who can not express their need in SQL.
- It can deal easily with the heterogeneity of sources.

(4) Disadvantages of semi-structured data

- Lack of fixed, rigid schema make it difficult in storage of the data.
- Interpreting the relationship between data is difficult as there is no separation of the schema and the data.
- Queries are less efficient as compared to structured data.

(5) Problems faced in storing semi-structured data

- Data usually has an irregular and partial structure. Some sources have implicit structure of data, which makes it difficult to interpret the relationship between data.
- Schema and data are usually tightly coupled, they are not only linked together but are also dependent of each other. Same query may update both schema and data with the schema being updated frequently.
- Distinction between schema and data is very uncertain or unclear. This complicates the designing of structure of data.
- Storage cost is high compared with structured data.

(6) Possible solutions for storing semi-structured data

- Data can be stored in DBMS specially designed to store semi-structured data.
- XML is widely used to store and exchange semi-structured data. It allows its user to define tags and attributes to store the data in a hierarchical form. Schema and data are not tightly coupled in XML.
- Object Exchange Model (OEM) can be used to store and exchange semi-structured data. OEM structures data in the form of graph.
- RDBMS can be used to store the data by mapping the data to relational schema and then mapping it to a table.

(7) Extracting information from semi-structured data

Semi-structured data has different structure because of heterogeneity of the sources. Sometimes it does not contain any structure at all. This makes it difficult to tag and index. So extracting information from it is a tough job. Here are some possible solutions.

- Graph-based models (e.g OEM) can be used to index semi-structured data.
- Data modelling technique in OEM allows the data to be stored in graph-based model. The data in graph-based model is easier to search and index.
- XML allows data to be arranged in hierarchical order which enables the data to be indexed and searched.
- Various data mining tools are used.

## New Words

| | | |
|---|---|---|
| structure | ['strʌktʃə] | n.结构；构造；体系<br>vt.构成，排列；安排 |
| consistent | [kən'sɪstənt] | adj.一致的；连续的 |
| schema | ['ski:mə] | n.概要，纲要；图表 |
| attribute | [ə'trɪbju:t]<br>['ætrɪbju:t] | v.把……归因于<br>n.属性；性质；特征 |
| row | [rəʊ] | n.行，排 |
| column | ['kɒləm] | n.列，纵队 |
| identifiable | [aɪ,dentɪ'faɪəbl] | adj.可识别的，可辨认的 |
| field | [fi:ld] | n.字段，域 |
| file | [faɪl] | n.文件 |
| entity | ['entəti] | n.实体 |
| relation | [rɪ'leɪʃn] | n.关系；联系；关联 |
| class | [klɑ:s] | n.类 |
| access | ['ækses] | vt.访问，存取（计算机信息） |
| query | ['kwɪəri] | v.查询<br>n.疑问，询问 |
| tag | [tæg] | n.标签<br>vt.加标签于 |
| network | ['netwɜ:k] | n.（计算机）网络 |
| server | ['sɜ:və] | n.服务器 |
| log | [lɒg] | n.记录；日志 |
| device | [dɪ'vaɪs] | n.装置，设备 |
| index | ['ɪndeks] | n.索引；指数<br>vt.给……编索引；将……编入索引 |
| string | [strɪŋ] | n.字符串 |

| search | [sɜ:tʃ] | *v.&n.*搜索；调查 |
| operation | [ˌɒpə'reɪʃn] | *n.*操作；运算 |
| undertake | [ˌʌndə'teɪk] | *vt.*承担，从事；同意，答应 |
| scalable | ['skeɪləbl] | *adj.*可伸缩的，可扩展的，可升级的 |
| ensure | [ɪn'ʃʊə] | *vt.*确保 |
| foundation | [faʊn'deɪʃn] | *n.*基础 |
| pre-defined | [pri:dɪ'faɪnd] | *adj.*预先定义的 |
| mainstream | ['meɪnstri:m] | *n.*主流<br>*adj.*主流的 |
| semantic | [sɪ'mæntɪk] | *adj.*语义的，语义学的 |
| rule | [ru:l] | *n.*规则<br>*v.*控制 |
| format | ['fɔ:mæt] | *n.*格式<br>*vt.*使格式化 |
| sequence | ['si:kwəns] | *n.*序列；顺序；连续<br>*vt.*安排顺序 |
| presentation | [ˌprezn'teɪʃn] | *n.*幻灯片 |
| flexible | ['fleksəbl] | *adj.*灵活的 |
| absence | ['æbsəns] | *n.*缺乏，缺少 |
| portable | ['pɔ:təbl] | *adj.*手提的；轻便的 |
| heterogeneity | [ˌhetərədʒə'ni:əti] | *n.*异质性，不均匀性 |
| unclear | [ˌʌn'klɪə] | *adj.*不清晰的，含糊不清的 |
| metadata | ['metədeɪtə] | *n.*元数据 |
| object | ['ɒbdʒɪkt] | *n.*物体；目标；对象 |
| retrieve | [rɪ'tri:v] | *vt.*检索；重新得到 |
| algorithm | ['ælgərɪðəm] | *n.*算法 |
| taxonomy | [tæk'sɒnəmi] | *n.*分类学，分类系统 |
| classification | [ˌklæsɪfɪ'keɪʃn] | *n.*分类，归类 |
| hierarchical | [ˌhaɪə'rɑ:kɪkl] | *adj.*分层的 |
| organizational | [ˌɔ:gənaɪ'zeɪʃnl] | *adj.*组织的 |
| property | ['prɒpəti] | *n.*特性；属性 |
| element | ['elɪmənt] | *n.*元素；要素；原理 |
| describe | [dɪ'skraɪb] | *vt.*叙述；描绘 |
| sufficient | [sə'fɪʃnt] | *adj.*足够的，充足的，充分的 |
| automation | [ˌɔ:tə'meɪʃn] | *n.*自动化（技术），自动操作 |
| binary | ['baɪnəri] | *adj.*二进制的；双重的；二元的 |
| executable | [ɪg'zekjətəbl] | *adj.*可执行的；实行的 |

| packet | ['pækɪt] | n.信息包 |
| | | vt.包装，打包 |
| rigid | ['rɪdʒɪd] | adj.严格的 |
| separation | [ˌsepə'reɪʃn] | n.分离，分开；间隔 |
| irregular | [ɪ'regjələ] | adj.不规则的；无规律的；不合规范的 |
| partial | ['pɑ:ʃl] | adj.部分的 |
| implicit | [ɪm'plɪsɪt] | adj.无疑问的，绝对的；成为一部分的 |
| tightly | ['taɪtli] | adv.紧紧地，坚固地，牢固地 |
| uncertain | [ʌn's3:tn] | adj.不确定的；不稳定的；不明确的 |
| complicate | ['kɒmplɪkeɪt] | vt.使复杂化；使错综，使混乱 |
| exchange | [ɪks'tʃeɪndʒ] | n.&v.交换，互换 |
| define | [dɪ'faɪn] | vi.下定义，构成释义 |
| | | n.定义 |
| graph | [grɑ:f] | n.图表，曲线图 |
| table | ['teɪbl] | n.表格 |
| | | vt.提交（议案） |

## ✍ Phrases

| structured data | 结构化数据 |
| unstructured data | 非结构化数据 |
| semi-structured data | 半结构化数据 |
| conform to | 符合，遵照 |
| computer program | 计算机程序 |
| data model | 数据模型 |
| base on | 基于 |
| business intelligence | 商业智能 |
| relational database | 关系数据库，关系型数据库 |
| web page | 网页 |
| a variety of | 多种的，各种各样的 |
| be converted to | 转换为……，改变为…… |
| reside in | 住在……，存在于…… |
| markup language | 标记语言，标识语言 |
| zipped file | 压缩文件 |
| be constrained by | 受……约束 |
| storage cost | 存储成本，存储花费 |

| | |
|---|---|
| be compared with | 与……相比较 |
| graph-based model | 基于图形模式 |

## ✎ Abbreviations

| | |
|---|---|
| SQL (Structured Query Language) | 结构化查询语言 |
| OLTP (On-Line Transaction Processing) | 联机事务处理 |
| GPS (Global Position System) | 全球定位系统 |
| RFID (Radio Frequency IDentification) | 射频识别 |
| JPEG(Joint Photographic Experts Group) | 联合图像专家组 |
| GIF (Graphics Interchange Format) | 图像互换格式 |
| PNG (Portable Network Graphics) | 便携式网络图形 |
| CAS (Content Addressable Storage) | 内容可寻址存储 |
| XML (eXtensible Markup Language) | 可扩展标记语言 |
| RDBMS (Relational DataBase Management System) | 关系数据库管理系统 |
| BLOB (Binary Large OBject) | 二进制长对象 |
| XOLAP (eXtended On-Line Analytic Processing) | 扩展联机分析处理 |
| TCP/IP (Transmission Control Protocol/Internet Protocol) | 传输控制协议/网际协议 |
| DBMS (DataBase Management System) | 数据库管理系统 |
| OEM (Object Exchange Model) | 对象交换模型 |

# Reference Translation

## 结构化数据、非结构化数据和半结构化数据

### 1. 结构化数据

结构化数据是符合数据模型、具有定义明确的结构、遵循一致顺序并且可以由人或计算机程序轻松访问和使用的数据。它通常存储在定义明确的架构（如数据库）中。它通常是表格形式，具有明确定义其属性的列和行。SQL 通常用于管理存储在数据库中的结构化数据。

（1）结构化数据的特征

● 数据符合数据模型，并具有易于识别的结构。

● 数据以行和列的形式存储。

● 数据井井有条，因此具有明确的定义、格式和含义。

● 数据位于记录或文件的固定字段中。

● 将相似的实体组合在一起以形成关系或类。

● 同一组中的实体具有相同的属性。

● 数据易于访问和查询，因此其他程序可以轻松使用。

（2）结构化数据的来源

● SQL 数据库。

- 电子表格，如 Excel。
- OLTP 系统。
- 在线表格。
- 传感器，如 GPS 或 RFID 标签。
- 网络和 Web 服务器日志。
- 医疗设备。

（3）结构化数据的优势

- 结构化数据具有定义明确的结构，有助于轻松存储和访问数据。
- 可以基于文本字符串和属性为数据建立索引，这使搜索操作变得轻松自如。
- 数据挖掘很容易，可以轻松地从数据中提取知识。
- 由于具有结构良好的数据形式，因此更新和删除等操作很容易。
- 可以轻松进行诸如数据仓库之类的商业智能操作。
- 如果数据有增加，则可以轻松扩展。
- 很容易确保数据的安全性。

注意：结构化数据仅占数据的20%左右，但其高度的组织性和性能使其成为大数据的基础。

## 2. 非结构化数据

非结构化数据是指不符合数据模型并且没有易于识别的结构的数据，因此不易被计算机程序所使用。非结构化数据不是以预定义的方式组织的，也不具有预定义的数据模型，因此它不适用于主流的关系数据库。

（1）非结构化数据的特征

- 数据既不符合数据模型也不具有任何结构。
- 数据不能以行和列的形式存储。
- 数据不遵循任何语义或规则。
- 数据缺少任何特定的格式或序列。
- 数据没有易于识别的结构。
- 由于缺乏可识别的结构，不易被计算机程序所使用。

（2）非结构化数据的来源

- 网页。
- 图像（JPEG、GIF、PNG 等）。
- 视频。
- 备忘录。
- 报告。
- Word 文档和 PowerPoint 演示文件。
- 调查。

（3）非结构化数据的优势

- 支持缺少正确格式或序列的数据。
- 数据不受固定模式的约束。
- 由于没有模式，因此非常灵活。
- 数据是可移植的。

- 它具有很好的可扩展性。
- 它可以轻松处理数据源的异构性。
- 它可以应用于各种商业智能和分析。

（4）非结构化数据的缺点

- 由于缺乏模式和结构，非结构化数据很难存储和管理。
- 索引数据很困难，并且由于结构不清晰且没有预定义的属性而导致错误，因此搜索结果不太准确。
- 确保数据的安全性是一项艰巨的任务。

（5）存储非结构化数据时面临的问题

- 需要大量存储空间来存储非结构化数据。
- 很难存储视频、图像、音频等。
- 由于结构不清楚，因此更新、删除和搜索等操作非常困难。
- 与结构化数据相比，存储成本很高。
- 索引非结构化数据很困难。

（6）用于存储非结构化数据的可能解决方案

- 非结构化数据可以转换为易于管理的格式。
- 使用内容可寻址存储（CAS）系统存储非结构化数据。
- 根据其元数据存储数据，并且为存储在其中的每个对象分配唯一的名称。根据内容而不是其位置来检索对象。
- 非结构化数据可以以 XML 格式存储。
- 非结构化数据可以存储在支持 BLOB 的 RDBMS 中。

（7）从非结构化数据中提取信息

非结构化数据没有任何结构。因此，传统算法不能轻易解释它。标记和索引非结构化数据也很困难。因此，从中提取信息是一项艰巨的工作。这里是一些可能的解决方案。

- 分类法或数据分类有助于按层次结构组织数据，这将使搜索过程变得容易。
- 数据可以存储在虚拟存储库中并被自动标记。
- 使用 XOLAP 之类的应用程序平台。XOLAP 帮助用户从电子邮件和基于 XML 的文档中提取信息。
- 使用各种数据挖掘工具。

## 3. 半结构化数据

半结构化数据是不符合数据模型但具有某种结构的数据。它缺乏固定或严格的模式。数据不是驻留在合理数据库中，但具有一些使其更易于分析的组织属性。通过一些方法，我们可以将它们存储在关系数据库中。

（1）半结构化数据的特征

- 数据不符合数据模型，但具有某种结构。
- 不能以行和列的形式存储数据。
- 半结构化数据包含标签和元素（元数据），这些标签和元素可用来对数据进行分组并描述数据的存储方式。
- 将相似的实体组合在一起并按层次结构进行组织。

- 同一组中的实体可能有也可能没有相同的属性。
- 它没有足够的元数据，这使得数据的自动化和管理变得困难。
- 一组中属性相同的数据其大小和类型可能不同。
- 由于缺乏明确定义的结构，它无法容易地被计算机程序使用。

（2）半结构化数据的来源

- 电子邮件。
- XML 和其他标记语言。
- 二进制可执行文件。
- TCP/IP 数据包。
- 压缩文件。
- 不同来源的整合数据。
- 网页。

（3）半结构化数据的优势

- 数据不受固定模式的约束。
- 灵活，即可以轻松更改架构。
- 数据是可移植的。
- 可以将结构化数据视为半结构化数据。
- 它为无法用 SQL 表达需求的用户提供支持。
- 它可以轻松处理数据源的异构性。

（4）半结构化数据的缺点

- 缺乏固定的和严格的架构，因此存储数据困难。
- 因为架构和数据没有分离，很难解释数据之间的关系。
- 与结构化数据相比，查询效率较低。

（5）存储半结构化数据时面临的问题

- 数据通常具有不规则的部分结构。一些数据源具有隐式的数据结构，这使得解释数据之间的关系很困难。
- 模式和数据通常紧密耦合，它们不仅链接在一起，而且彼此依赖。同一查询可能会同时更新架构和数据，而架构会经常更新。
- 模式和数据之间的区别非常不确定或不清楚，这使数据结构的设计复杂化。
- 与结构化数据相比，存储成本很高。

（6）存储半结构化数据的可能解决方案

- 数据可以存储在专门为存储半结构化数据而设计的 DBMS 中。
- XML 被广泛用于存储和交换半结构化数据。它允许用户定义标签和属性，以分层形式存储数据。模式和数据在 XML 中并没有紧密耦合。
- 对象交换模型（OEM）可用于存储和交换半结构化数据。OEM 以图形形式构造数据。
- RDBMS 可以通过将数据映射到关系模式，然后将其映射到表来存储数据。

（7）从半结构化数据中提取信息

由于数据源的异构性，半结构化数据具有不同的结构。有时它根本不包含任何结构。这使标记和索引数据很困难。因此，从中提取信息是一项艰巨的工作。这里是一些可能的解决方案。

- 基于图形模型（如 OEM）可用于索引半结构化数据。
- OEM 中的数据建模技术允许将数据存储在基于图形的模型中。基于图形的模型中的数据更易于搜索和索引。
- XML 允许按层次结构排列数据，从而可以对数据进行索引和搜索。
- 使用各种数据挖掘工具。

# Text B

## Data Modeling

扫码听课文

Data modeling is the process of documenting a complex software system design as an easily understood diagram. The diagram uses text and symbols to represent the way data needs to flow. The diagram can be used to ensure efficient use of data as a blueprint for the construction of new software or for re-engineering a legacy application.

Data modeling is an important skill for data scientists or others involved with data analysis. Traditionally, data models have been built during the analysis and design phases of a project to ensure that the requirements for a new application are fully understood. Data models can also be invoked later in the data lifecycle to rationalize data designs that were originally created by programmers on an ad hoc basis.

### 1. Data modeling approaches

Data modeling can be a painstaking upfront process and is sometimes seen as being at odds with rapid development methodologies. As agile programming has come into wider use to speed development projects, after-the-fact methods of data modeling are being adapted in some instances. Typically, a data model can be thought of as a flowchart that illustrates the relationships among data. It enables stakeholders to identify errors and make changes before any code has been written. Alternatively, models can be introduced as part of reverse engineering efforts that extract models from existing systems,such as NoSQL data.

Data modelers often use multiple models to view the same data and ensure that all processes, entities, relationships and data flows have been identified. They initiate new projects by gathering requirements from business stakeholders. Data modeling stages roughly break down into creation of logical data models that show specific attributes, entities and relationships among entities and the physical data model.

The logical data model serves is the basis for creation of a physical data model, which is specific to the application and database to be implemented. A data model can become the basis for building a more detailed data schema.

### 2. Hierarchical data modeling

Data modeling as a discipline began to arise in the 1960s, accompanying the upswing in use of database management systems (DBMS). Data modeling enabled organizations to bring consistency, repeatability and well-ordered development to data processing. Application end users and programmers were able to use the data model as a reference in communications with data designers.

Hierarchical data models that array data in treelike, one-to-many arrangements marked these early efforts and replaced file-based systems in many popular use cases. IBM's Information Management System (IMS) is a primary example of the hierarchical approach, which found wide use in businesses, especially in banking. Although hierarchical data models were largely replaced by the relational data models starting in the 1980s, the hierarchical method is common still in XML and Geographic Information Systems (GIS) today. Network data models also arose in the early days of DBMS as a means to provide data designers with a broad conceptual view of their systems. One such example is the Conference on Data Systems Languages (CODASYL), which formed in the late 1950s to guide the development of a standard programming language that could be used across various types of computers.

## 3. Relational data modeling

While it reduced program complexity versus file-based systems, the hierarchical model still required detailed understanding of the specific physical data storage employed. Proposed as an alternative to the hierarchical data model, the relational data model does not require developers to define data paths. Relational data modeling was first described in a 1970 technical paper by IBM researcher E. F. Codd. Codd's relational model set the stage for industry use of relational databases, in which data segments were explicitly joined by use of tables, as compared to the hierarchical model where data was implicitly joined together. Soon after its inception, the relational data model was coupled with the Structured Query Language (SQL) and began to gain an ever larger foothold in enterprise computing as an efficient means to process data.

## 4. The entity-relationship model

Relational data modeling took another step forward in the mid-1970s as use of Entity-Relationship (ER) models became more prevalent. Closely integrated with relational data models, ER models use diagrams to graphically depict the elements in a database and to ease understanding of underlying models.

With relational modeling, data types are determined and rarely change over time. Entities comprise attributes, for example, an employee entity's attributes could include last name, first name, years employed and so on. Relationships are visually mapped, providing a ready means to communicate data design objectives to various participants in data development and maintenance. Over time, modeling tools, including Idera's ER/Studio, ERwin Data Modeler and SAP PowerDesigner, gained wide use among data architects for designing systems.

## 5. Object-oriented model

As object-oriented programming gained ground in the 1990s, object-oriented modeling gained wide attention as yet another way to design systems. While bearing some resemblance to ER methods, object-oriented approaches differ in that they focus on object abstractions of real-world entities. Objects are grouped in class hierarchies, and the objects within such class hierarchies can inherit attributes and methods from parent classes. Because of this inheritance trait, object-oriented data models have some advantages versus ER modeling, in terms of ensuring data integrity and supporting more complex data relationships. Also arising in the 1990s were data models specifically oriented toward data warehousing needs. Notable examples are snowflake schema and star schema dimensional models.

**6. Graph data models**

An offshoot of hierarchical and network data modeling is the property graph model, which, together with graph databases, has found increased use for describing complex relationships within data sets, particularly in social media, recommender and fraud detection applications.

Using the graph data model, designers describe their system as a connected graph of nodes and relationships, much as they might do with ER or object data modeling. Graph data models can be used for text analysis, creating models that uncover relationships among data points within documents.

## New Words

| | | |
|---|---|---|
| diagram | ['daɪəgræm] | n.图表；图解；示意图 |
| symbol | ['sɪmbl] | n.符号；标志 |
| blueprint | ['blu:prɪnt] | n.蓝图，设计图；计划大纲 |
| construction | [kən'strʌkʃn] | n.建造；建筑物 |
| re-engineering | ['ri:endʒɪ'nɪərɪŋ] | n.再设计，重建 |
| | | v.再造；重建 |
| legacy | ['legəsi] | n.遗产 |
| phase | [feɪz] | n.阶段 |
| lifecycle | ['laɪf,saɪkl] | n.生命周期 |
| programmer | ['prəugræmə] | n.程序员，程序设计者 |
| adhoc | [,æd'hɒk] | adj.特别的，特设的；临时的 |
| approach | [ə'prəutʃ] | n.方法；途径 |
| | | v.接近，走近，靠近 |
| painstaking | ['peɪnzteɪkɪŋ] | adj.艰苦的，辛苦的 |
| | | n.辛苦，勤勉 |
| instance | ['ɪnstəns] | n.实例 |
| | | v.举……为例 |
| implement | ['ɪmplɪment] | vt.实施，执行；实现 |
| | | n.工具，器械；手段 |
| consistency | [kən'sɪstənsi] | n.一致性 |
| repeatability | [rɪ'pi:tə'bɪlɪti] | n.可重复性；反复性；再现性 |
| conceptual | [kən'septʃuəl] | adj.观念的，概念的 |
| standard | ['stændəd] | n.标准，规格 |
| | | adj.标准的 |

| complexity | [kəm'pleksəti] | n.复杂性 |
| path | [pɑ:θ] | n.路径，路线 |
| inception | [ɪn'sepʃn] | n.开始，开端，初期 |
| depict | [dɪ'pɪkt] | vt.描绘，描述 |
| prevalent | ['prevələnt] | adj.流行的；普遍的 |
| visually | ['vɪʒuəli] | adv.可视化地；视觉上 |
| maintenance | ['meɪntənəns] | n.维护；维修 |
| resemblance | [rɪ'zembləns] | n.相似，相似之处 |
| inherit | [ɪn'herɪt] | v.继承 |
| dimensional | [dɪ'menʃənəl] | adj.维度的，维数的 |
| recommender | [ˌrekə'mendə] | n.推荐引擎；推荐系统 |
| node | [nəʊd] | n.节点 |

## ✍ Phrases

| data modeling | 数据建模 |
| be involved with | 与……有密切关系，涉及…… |
| rapid development methodology | 快速开发方法 |
| agile programming | 敏捷编程 |
| after-the-fact method | 事后方法 |
| reverse engineering | 逆向工程，反向工程 |
| logical data model | 逻辑数据模型 |
| physical data model | 物理数据模型 |
| serve as | 充当，担任 |
| hierarchical data model | 层次数据模型，分级数据模型 |
| file-based system | 基于文件的系统 |
| network data model | 网络数据模型 |
| relational data model | 关系数据模型 |
| data segment | 数据段 |
| be couple with | 与……一起，连同…… |
| entity-relationship model | 实体关系模型 |
| object-oriented modeling | 面向对象建模 |
| be grouped in | 用……分组，按……分组 |
| parent class | 父类 |
| snowflake schema | 雪花模式 |
| star schema | 星型模式 |
| graph data model | 图形数据模型 |
| data set | 数据集 |

| | |
|---|---|
| fraud detection | 欺诈检测 |
| text analysis | 文本分析 |

## ✍ Abbreviations

| | |
|---|---|
| IMS (Information Management System) | 信息管理系统 |
| GIS (Geographic Information System) | 地理信息系统 |
| CODASYL (Conference on Data Systems Languages) | 数据系统语言会议 |
| ER (Entity-Relationship) | 实体关系 |

## Reference Translation

# 数 据 建 模

数据建模是将复杂的软件系统设计记录为易于理解的图的过程，该图使用文本和符号来表示数据需求流动的方式。该图可用于确保有效利用数据，作为构建新软件或重新设计旧版应用程序的蓝图。

数据建模是数据科学家或其他参与数据分析的人员的一项重要技能。传统上，数据模型在项目的分析和设计阶段建立，以确保充分理解新应用程序的需求。数据模型也可以在数据生命周期的后期调用，以使最初由程序员临时创建的数据设计更为合理。

### 1. 数据建模方法

数据建模可能是一个艰苦的前期过程，有时被视为与快速开发方法不符。随着敏捷编程已被广泛用于加速开发项目，在某些情况下正在采用事后的数据建模方法。通常，数据模型可以看作是流程图，它说明了数据之间的关系。它使利益相关者能够在编写任何代码之前识别错误并进行更改。或者，可以将模型作为逆向工程工作的一部分引入，以从现有系统中提取模型，如 NoSQL 数据。

数据建模人员经常使用多个模型来查看相同的数据，并确保已识别所有流程、实体、关系和数据流。他们通过收集业务利益相关者的需求来启动新项目。数据建模阶段大致分成创建逻辑数据模型，这些逻辑数据模型显示特定的属性、实体以及实体与物理数据模型之间的关系。

逻辑数据模型是创建物理数据模型的基础，该物理数据模型针对要实现的应用程序和数据库。数据模型可以成为构建更详细的数据架构的基础。

### 2. 分层数据建模

随着数据库管理系统（DBMS）使用的兴起，数据建模作为一门学科于 20 世纪 60 年代开始出现。数据建模使组织能够将数据处理得具有一致性、可重复性和有序开发。应用程序的最终用户和程序员能够将数据模型用作与数据设计人员进行沟通的参考。

以树状、一对多的方式排列数据的分层数据模型标记了这些早期工作，并在许多流行的用例中取代了基于文件的系统。IBM 的信息管理系统（IMS）是分层方法的主要示例，该方法在企业（尤其是银行业务）中得到了广泛应用。尽管从 20 世纪 80 年代开始，分层数据模型在很大程度上被关系数据模型所取代，但是分层方法在当今的 XML（可扩展标记语言）和地理信息系统（GIS）中仍然很常见。在 DBMS 的早期，网络数据模型也应运而生，它是一种为数据设计人员提供其系统的广泛概念视图的方法。这样的例子之一是数据系统语言会议（CODASYL），该会议成立于 20 世纪 50 年代后期，旨在指导可在各种类型的计算机上使用的标准编程语言的开发。

## 3. 关系数据建模

与基于文件的系统相比，尽管它降低了程序的复杂性，但分层模型仍然需要详细了解所采用的特定物理数据存储。作为替代分层数据模型的方案，关系数据模型不需要开发人员定义数据路径。关系数据建模最早是在 1970 年由 IBM 研究人员 E. F. Codd 撰写的技术论文中描述的。Codd 的关系模型为行业使用关系数据库奠定了基础，与将数据隐式连接在一起的分层模型不同，它通过使用表格将数据段显式地连接在一起。关系数据模型一经问世便与结构化查询语言（SQL）结合使用，并开始在企业计算中日益稳固，成为处理数据的有效手段。

## 4. 实体关系模型

随着实体关系（ER）模型的使用变得越来越普遍，关系数据建模在 20 世纪 70 年代中期又向前迈出了一步。实体关系模型与关系数据模型紧密集成，使用图表以图形方式描述数据库中的元素，并简化对基础模型的理解。

通过关系建模，可以确定数据类型，并且数据类型很少随时间变化。实体包括属性，例如，员工实体的属性可以包括姓、名、受雇年限等。关系被可视化地映射，为将数据设计目标传达给数据开发和维护的各个参与者提供了现成的手段。随着时间的流逝，设计系统的数据架构师都在广泛使用包括 Idera 的 ER/Studio、ERwin Data Modeler 和 SAP PowerDesigner 在内的建模工具。

## 5. 面向对象模型

20 世纪 90 年代，随着面向对象编程的发展，面向对象建模作为设计系统的另一种方式受到了广泛的关注。面向对象的方法虽然与实体关系方法有些相似，但区别在于面向对象的方法专注于现实世界实体的对象抽象。对象按类层次结构分组，此类层次结构中的对象可以从父类继承属性和方法。由于具有这种继承特性，在确保数据完整性和支持更复杂的数据关系方面，面向对象的数据模型相对于实体关系建模具有一些优势。在 20 世纪 90 年代还出现了专门针对数据仓库需求的数据模型。值得注意的示例是雪花模式和星型维度模型。

## 6. 图形数据模型

层次图和网络数据建模的一个分支是属性图模型，它与图数据库一起已越来越多地用于描述数据集内的复杂关系，尤其是在社交媒体、推荐器和欺诈检测应用程序中。

设计人员使用图数据模型，将其系统描述为节点和关系的连接图，这与他们对实体关系或对象数据建模所做的工作很相似。图形数据模型可用于文本分析，创建可揭示文档内数据点之间关系的模型。

# Exercises

**[Ex. 1] 根据 Text A 填空。**

1. Structured data is the data which conforms to _____, has _____, follows _____ and can be easily accessed and used by a person or _____.

2. The advantages of structured data include data can be indexed based on _____ as well as _____. This makes search operation hassle-free. It is easily scalable in case there is _____.

3. Structured data accounts for only about _____ of data but its high degree of organization and performance make it _____.

4. Unstructured data is not organized in _____ nor has a _____, thus it is not a good fit for _____.

5. Unstructured data can not be stored in the form of _____. Due to lack of _____, it can not used by computer programs easily.

6. The disadvantages of unstructured data include it is difficult to _____ unstructured data due to _____. Indexing the data is difficult and error prone due to _____ and not having pre-defined attributes, due to which search results are _____.

7. Semi-structured data is the data which _____ a data model but has some structure. It lacks _____. It is the data that does not reside in _____ but that has _____ that make it easier to analyse.

8. Sources of semi-structured data are _____, XML and other markup languages, _____, _____, _____, Integration of data from different sources and _____.

9. XML is widely used to store and exchange _____. It allows its user to define _____ and _____ to store the data in _____. Schema and data are not tightly coupled in_____.

10. Semi-structured data has different structure because of _____. Sometimes it does not contain _____ at all. This makes it difficult to _____.

**[Ex. 2]** 根据 Text B 回答以下问题。

1. What is data mining?

2. What can a data model be thought of typically? What can it do?

3. What do data modeling stages roughly break down into?

4. How do hierarchical data models array data? What is a primary example of the hierarchical approach?

5. Where was relational data modeling first described? By whom?

6. What did Codd's relational model do?

7. What are ER models closely integrated with? What do ER models do?

8. What do object-oriented approaches do?

9. What is an offshoot of hierarchical and network data modeling?

10. What can graph data models be used for?

## [Ex. 3] 词汇英译中

| | |
|---|---|
| 1. data model | 1. _____ |
| 2. binary | 2. _____ |
| 3. unstructured data | 3. _____ |
| 4. class | 4. _____ |
| 5. data set | 5. _____ |
| 6. node | 6. _____ |
| 7. data segment | 7. _____ |
| 8. phase | 8. _____ |
| 9. hierarchical data model | 9. _____ |
| 10. symbol | 10. _____ |

## [Ex. 4] 词汇中译英

| | |
|---|---|
| 1. 快速开发方法 | 1. _____ |
| 2. 关系数据模型 | 2. _____ |
| 3. 操作；运算 | 3. _____ |
| 4. 检索；重新得到 | 4. _____ |
| 5. 文本分析 | 5. _____ |
| 6. 服务器 | 6. _____ |
| 7. 算法 | 7. _____ |
| 8. 概要，纲要；图表 | 8. _____ |
| 9. 属性；性质；特征 | 9. _____ |
| 10. 存储成本，存储花费 | 10. _____ |

## [Ex. 5] 短文翻译

### Metadata

#### 1. What is metadata

Quite simply: metadata is data that describes other data. In information technology, the prefix meta means "an underlying definition or description". So, metadata describes whatever piece of data it's connected to whether that data is video, photograph, web pages, content or spreadsheets.

Since metadata summarises basic info about data such as type of asset, author, date created, usage, file size and more, metadata is crucial to the efficiency of information systems to classify and categorise data. Metadata information helps IT systems uncover what users are looking for.

It's important to note that organizations are inundated with structured and unstructured data and they both need metadata. Structured data is easily organized and discovered through search engine algorithm (a strict database format), while unstructured data is the complete opposite. Email is an example of unstructured data. Most emails aren't easily categorised, because they rarely cover a single subject.

Most business interactions are in the format of unstructured data, making sorting and defining the data a time-consuming and expensive proposition, but metadata can help.

**2. Why does metadata matter in a big data world**

Metadata is a game-changer in the big data world, because it can give you a competitive advantage.

The better you harness the power of big data to drive business decisions, the more successful your firm will be. The more robust your metadata, the quicker your team will be able to extract actionable information and make quick business decisions. In addition to better and quicker decision-making, metadata supports data consistency across an enterprise and enables associations between data sets for high-quality results.

Although metadata is one of the fastest-growing segments of enterprise data management according to a report published by IDC, there's a significant Big Data Gap—metadata isn't keeping up with the rapid rate of big data projects. Without metadata, companies are losing out on analysing and interpreting big data and the subsequent insight it offers to propel their business.

# Reading Material

## Data Structure

A data structure[1] is a specialized format for organizing, processing, retrieving and storing data. While there are several basic and advanced structure types, any data structure is designed to arrange data to suit a specific purpose so that it can be accessed and worked with in appropriate ways.

In computer programming, a data structure may be selected or designed to store data for the purpose of working on it with various algorithms. Each data structure contains information about the data values, relationships between the data and functions that can be applied to the data.

**1. Characteristics of data structures**

Data structures are often classified by their characteristics. Possible characteristics are as follows.

- Linear or non-linear: This characteristic describes whether the data items are arranged in a chronological[2] sequence, such as with an array, or in an unordered sequence, such as with a graph.

---

1　data structure: 数据结构
2　chronological: [ˌkrɒnəˈlɒdʒɪkəl] *adj.* 按时间的前后顺序排列的

- Homogeneous[3] or non-homogeneous: This characteristic describes whether all data items in a given repository are of the same type or of various types.
- Static[4] or dynamic[5]: This characteristic describes how the data structures are compiled. Static data structures have fixed sizes, structures and memory locations at compile time. Dynamic data structures have sizes, structures and memory locations that can shrink or expand depending on the use.

## 2. Types of data structures

Data structures types are determined by what types of operations are required or what kinds of algorithms are going to be applied. These types include:

- Arrays[6]. An array stores a collection of items at adjoining memory locations. Items that are of the same type get stored together so that the position of each element can be calculated or retrieved easily. Arrays can be fixed or flexible in length.
- Stacks[7]. A stack stores a collection of items in the linear order that operations are applied. This order could be Last in First out[8] (LIFO).
- Queues[9]. A queue stores a collection of items similar to a stack; however, the operation order can only be first in first out.
- Linked lists[10]. A linked list stores a collection of items in a linear order. Each element, or node, in a linked list contains a data item as well as a reference, or link, to the next item in the list.
- Trees. A tree stores a collection of items in an abstract, hierarchical way. Each node is linked to other nodes and can have multiple sub-values, also known as children.
- Graphs. A graph stores a collection of items in a non-linear fashion. Graphs are made up of a finite set of nodes, also known as vertices and lines that connect them, also known as edges. These are useful for representing real-life systems such as computer networks.
- Tries. A trie, or keyword[11] tree, is a data structure that stores strings as data items that can be organized in a visual graph.
- Hash tables[12]. A hash table, or a hash map, stores a collection of items in an associative[13] array that plots keys to values. A hash table uses a hash function to convert an index into an array of buckets that contain the desired data item.

---

3  homogeneous [ˌhɒməˈdʒiːnɪəs] *adj.*同性质的，同类的

4  static[ˈstætɪk] *adj.*静态的

5  dynamic [daɪˈnæmɪk] *adj.*动态的

6  array [əˈreɪ] *n.*数组；数列

7  stack [stæk] *n.*堆栈

8  last in first out: 后进先出

9  queue [kjuː] *n.*队列

10  linked list: 链表

11  keyword [ˈkiːwɜːd] *n.*关键字

12  hash table: 哈希表

13  associative [əˈsəʊʃɪətɪv] *adj.*关联的，联合的

These are considered complex data structures as they can store large amounts of interconnected data. Examples of primitive, or basic, data structures are integers, floats[14], Booleans and characters[15].

## 3. Uses of data structures

In general, data structures are used to implement the physical forms of abstract data types. This can be translated into a variety of applications, such as displaying a relational database as a binary tree[16].

In programming languages, data structures are used to organize code and information in a digital space. For example, Python lists and dictionaries or JavaScript arrays and objects are common coding structures used for storing and retrieving information. Data structures are also a crucial part of designing efficient software.

## 4. Importance of data structures

Data structures are essential for efficiently managing large amounts of data, such as information kept in databases or indexing services. Proper maintenance of data systems requires the identification of memory allocation, data interrelationships and data processes.

Additionally, it is not only important to use data structures but it is important to choose the proper data structure for each task. Choosing an ill-suited data structure could result in slow runtimes or unresponsive[17] codes. A few factors to consider when picking a data structure include what kind of information will be stored, where should existing data be placed, how should data be sorted and how much memory should be reserved for the data.

---

14　float [fləʊt] *n.*浮点数

15　character ['kærəktə] *n.*字符

16　binary tree: 二叉树

17　unresponsive [ˌʌnrɪ'spɒnsɪv] *adj.*无响应的

# Unit 3

## Text A

扫码听课文

### Data Collection

**1. How to collect data**

(1) Determine what information you want to collect

The first thing you need to do is to choose what details you want to collect. You'll need to decide what topics the information will cover, who you want to collect it from and how much data you need. Your goals (what you hope to accomplish using your data) will determine your answers to these questions. As an example, you may decide to collect data about which type of articles are most popular on website among visitors who are between the ages of 18 and 34. You might also choose to gather information about the average age of all of the customers who bought a product from your company within the last month.

(2) Set a timeframe for data collection

Next, you can start formulating your plan for how to collect your data. In the early stages of your planning process, you should establish a timeframe for your data collection. You may want to gather some types of data continuously. When it comes to transactional data and website visitor data, for example, you may want to set up a method for tracking that data over the long term. If you're tracking data for a specific campaign, however, you'll track it over a defined period. In these instances, you should have a schedule for when you'll start and end your data collection.

(3) Determine data collection method

At this step, you will choose the data collection method that will make up the core of your data-gathering strategy. To select the right collection method, you'll need to consider the type of information you want to collect, the timeframe over which you'll obtain it and the other aspects you are determined.

(4) Collect the data

Once you have finalized your plan, you can implement your data collection strategy and start collecting data. You can store and organize your data in your DMP (Data Management Platform). Be sure to stick to your plan and check on its progress regularly. It may be useful to create a schedule for when you will check in with how your data collection is proceeding, especially if you are collecting data continuously. You may want to make updates to your plan as conditions change and you get new information.

## 2. Ways to collect data

So, how do you go about collecting the data you need to meet your goals? There are various methods of collecting primary, quantitative data. Some involve directly asking customers for information, some involve monitoring your interactions with customers and others involve observing customers' behavior. The right one depends on your goals and the type of data you're collecting. Here are some of the most common types of data collection used today.

(1) Surveys

Surveys are one way in which you can directly ask customers for information. You can use them to collect either quantitative or qualitative data or both. A survey consists of a list of queries respondents can answer in just one or two words and often gives participants a list of responses to choose from. You can conduct surveys online, over email, over the phone or in person. One of the easiest methods is to create an online survey you host on your website or with a third party. You can then share a link to that survey on social media, over email and in pop-ups on your web site.

(2) Online tracking

Your business website and your app if you have one, are excellent tools for collecting customer data. When someone visits your website, they create as many as 40 data points. Accessing this data allows you to see how many people visited your website, how long they were on it, what they clicked on and more. Your website hosting provider may collect this kind of information, and you can also use analytics software. You can also place and read cookies to help track user behavior. Lotame can help you with this online data collection process.

(3) Transactional data tracking

Whether you sell goods in-store, online or both, your transactional data can give you valuable insights about your customers and your business. You may store transactional records in a customer relationship management system. That data may come from your web store, a third party you contract with for e-commerce or your in-store point-of-sale system. This information can give you insights about how many products you sell, what types of products are most popular, how often people typically purchase from you and more.

(4) Online marketing analytics

You can also collect valuable data through your marketing campaigns, whether you run them on search, web pages, email or elsewhere. You can even import information from offline marketing campaigns that you run. The software you use to place your ads will likely give you data about who clicked on your ads, what time they clicked, what device they used and more. Lotame Insights can also help you gather data about your campaigns. If you track the performance of offline ads by, for example, asking customers how they heard about your brand, you can import that data into your DMP.

(5) Social media monitoring

Social media is another excellent source of customer data. You can look through your follower list to see who follows you and what characteristics they have in common to enhance your understanding of who your target audience should be. You can also monitor mentions of your brand on social media by regularly searching your brand's name, setting up alerts or using third-party social media monitoring software. Many social media websites will also provide you with analytics about how your posts perform. Third-party tools may be able to offer you even more in-depth insights.

(6) Collecting subscription and registration data

Offering customers something in return for providing information about themselves can help you gather valuable customer data. You can do this by requiring some basic information from customers or website visitors who want to sign up for your email list, rewards program or another similar program. One benefit of this method is that the leads you get are likely to convert because they have actively demonstrated an interest in your brand. When creating the forms used to collect this information, it's essential to find the right balance in the amount of data you ask for. Asking for too much may make people not want to participate, while not asking for enough means your data won't be as useful as it could be.

(7) In-store traffic monitoring

If you have a brick-and-mortar store, you can also gather insights from monitoring the foot traffic there. The most straightforward way to do this is with a traffic counter on the door to help you keep track of how many people come into your store throughout the day. This data will reveal your busiest days and hours. It may also help give you an idea about what is drawing customers to your store at certain times. You can also install security systems with motion sensors that will help you track customers' movement patterns throughout your shop. The sensor can provide you with data about which of your store's departments are most popular.

## New Words

| | | |
|---|---|---|
| collect | [kə'lekt] | vt.收集 |
| website | ['websaɪt] | n.网站 |
| timeframe | ['taɪmfreɪm] | n.时间表 |
| stage | [steɪdʒ] | n.阶段 |
| establish | [ɪ'stæblɪʃ] | vt.建立，创建，确立 |
| continuously | [kən'tɪnjuəsli] | adv.连续不断地 |
| transactional | [træn'zækʃənəl] | adj.交易的，业务的 |
| tracking | ['trækɪŋ] | n.跟踪 |
| campaign | [kæm'peɪn] | n.活动；运动 |
| schedule | ['ʃedjuːl] | n.时刻表，进度表；清单，明细表<br>vt.排定，安排 |
| data-gathering | ['deɪtə 'gæðərɪŋ] | n.数据收集 |
| regularly | ['regjələli] | adv.有规律地，按时；整齐地；不断地；定期地 |
| quantitative | ['kwɒntɪtətɪv] | adj.定量的；数量（上）的 |

| monitoring | ['mɒnɪtərɪŋ] | n.监视；控制；监测 |
|---|---|---|
| interaction | [,ɪntər'ækʃn] | n.互相影响；互动 |
| behavior | [bɪ'heɪvjə] | n.行为；态度 |
| qualitative | ['kwɒlɪtətɪv] | adj.定性的 |
| participant | [pɑ:'tɪsɪpənt] | n.参加者，参与者 |
| | | adj.参与的 |
| online | [,ɒn'laɪn] | adj.在线的；联网的；联机的 |
| share | [ʃeə] | vi.分享；共有 |
| link | [lɪŋk] | v.连接 |
| | | n.联系，关系 |
| pop-up | ['pɒpʌp] | adj.弹出的 |
| excellent | ['eksələnt] | adj.卓越的；杰出的；优秀的 |
| click | [klɪk] | n&v.点击 |
| follower | ['fɒləuə] | n.关注者，追随者，拥护者 |
| mention | ['menʃn] | vt.提到，说起 |
| brand | [brænd] | n.商标，品牌 |
| subscription | [səb'skrɪpʃn] | n.（报刊等的）订阅费；（俱乐部等的）会员费 |
| registration | [,redʒɪ'streɪʃn] | n.登记，注册 |
| demonstrate | ['demənstreɪt] | vt.证明，证实；显示，展示 |
| essential | [ɪ'senʃl] | adj.基本的；必要的；本质的 |
| balance | ['bæləns] | n.平衡 |
| straightforward | [,streɪt'fɔ:wəd] | adj.直截了当的；坦率的；明确的 |
| counter | ['kauntə] | n.计数器 |
| sensor | ['sensə] | n.传感器 |

## ✎ Phrases

| data collection | 数据收集 |
|---|---|
| set up | 建立，设立；安排 |
| make up | 组成；补足 |
| consist of | 包含；由……组成 |
| third party | 第三方 |
| data point | 数据点 |
| customer relationship management system | 客户关系管理系统 |
| target audience | 目标受众 |
| sign up for | 报名；注册 |
| brick-and-mortar store | 实体店 |

## ✎ Abbreviations

DMP (Data Management Platform)                    数据管理平台

# Reference Translation

# 数 据 收 集

**1. 如何收集数据**

（1）确定要收集的信息

收集数据需要做的第一件事是选择要收集的详细信息。你需要确定信息将涉及哪些主题、希望从谁那里收集信息以及需要多少数据。你的目标（即希望使用数据完成的目标）将决定这些问题的答案。例如，你可能决定收集在网站上哪种类型的文章最受 18～34 岁的访问者欢迎。你还可以收集上个月内从你的公司购买了产品的客户平均年龄的信息。

（2）设置数据收集时间表

接下来，你可以开始制订如何收集数据的计划。在计划过程的早期阶段，你应该为数据收集建立时间表。你可能需要连续地收集某些类型的数据。例如，当涉及交易数据和网站访问者数据时，你可能希望建立一种长期跟踪该数据的方法。但是，如果你要跟踪特定活动的数据，则会在定义的时间内对其进行跟踪。在这些情况下，你应制定数据收集开始和结束的时间表。

（3）确定数据收集方法

在此步骤中，你将选择构成数据收集策略核心的数据收集方法。要选择正确的收集方式，需要考虑要收集的信息类型、获取信息的时间范围以及确定其他方面。

（4）收集数据

最终确定计划后，你可以实施数据收集策略并开始收集数据。你可以在 DMP（数据管理平台）中存储和组织数据。确保遵守你的计划并定期检查其进度。创建一个计划，确定检查数据收集进展情况的时间，尤其是连续收集数据时，这可能会很有用。随着条件的变化和获得新信息，你可能希望对计划进行更新。

**2. 数据收集方式**

因此，你如何收集实现目标所需的数据？有多种收集原始、定量数据的方法。有些涉及直接向客户询问信息，有些涉及监管你与客户的互动，而其他涉及观察客户的行为。合适的方法取决于你的目标和收集的数据类型。以下是当今使用的一些最常见的数据收集类型。

（1）调查

调查是你可以直接请求客户提供信息的一种方式。可以使用它们来收集定量或定性数据，或两者兼而有之。一项调查由受访者仅用一个或两个词回答的查询列表组成，通常会为参与者提供可供选择的答案列表。你可以在线、通过电子邮件、通过电话或亲临现场进行调查。一种最简单的方法是在你的网站或第三方托管的网站上创建在线调查。然后，你可以在社交媒体上、通过电子邮件和网站上的弹出窗口共享该调查的链接。

（2）在线跟踪

你的公司的网站以及你的应用程序（如果你有的话）都是收集客户数据的绝佳工具。有人访问你的网站时，他们会创建多达 40 个数据点。访问此数据，你可以查看有多少人访问过你的网站、他们在网站上停留了多长时间、点击了什么等。你的网站托管提供商可能会收集此类信息，并且你也可以使用分析软件。你还可以放置和读取 Cookie 来帮助跟踪用户行为。Lotame 可以帮助你完成此在线数据收集过程。

（3）交易数据跟踪

无论你是在实体店、在线商店销售商品，还是两者都有，交易数据都可以为你提供有关客户和业务的宝贵见解。你可以将交易记录存储在客户关系管理系统中。这些数据可能来自你的网上商店、与你签约进行电子商务的第三方或你的店内销售点系统。这些信息可以为你提供有关销售多少产品、最受欢迎的产品类型、人们通常从你那里购买商品的频率以及更多的见解。

（4）在线营销分析

你还可以通过市场营销活动收集有价值的数据，无论是在搜索、网页、电子邮件还是其他任何地方运行它们。你甚至可以从你运行的离线营销活动中导入信息。你用来放置广告的软件可能会为你提供有关谁点击了你的广告、他们什么时间点击的、他们使用什么设备等数据。Lotame Insights 还可以帮助你收集营销活动的数据。例如，如果想通过询问客户了解你的品牌的方式来跟踪离线广告的效果，则可以将该数据导入 DMP。

（5）社交媒体监控

社交媒体是另一个很好的客户数据来源。你可以浏览你的关注者列表，以了解谁关注你以及他们的共同特征，以加深你对目标受众的了解。还可以通过定期搜索品牌名称、设置提醒或使用第三方社交媒体监控软件来监控社交媒体上提及你的品牌。许多社交媒体网站还为你提供有关帖子效果的分析。第三方工具也许能够为你提供更深入的见解。

（6）收集订阅和注册数据

给客户一些东西，让他们提供自身的信息，可以帮助你收集有价值的客户数据。你可以要求那些要注册电子邮件列表、奖励计划或其他类似计划的客户或网站访问者提供一些基本信息。这种方法的一个好处是，你获得的潜在客户很可能会转化，因为他们已经积极展示了对你的品牌的兴趣。创建用于收集此信息的表格时，必须在所需的数据量中找到适当的平衡。索取过多可能会让人们不想参与，而索取不足意味着你的数据将无法发挥应有的作用。

（7）店内流量监控

如果你有实体商店，还可以通过监控那里的客流量来收集见解。最简单的方法是在门上设置一个客流量计数器，以帮助你跟踪一天中有多少人进入你的商店。此数据将显示你最忙的日子和时间。它还可以帮助你了解在特定时间能吸引顾客到你商店的原因。你还可以安装带有运动传感器的安全系统，以帮助你在整个商店中跟踪客户的移动模式。该传感器可以为你提供商店里哪个部门最受欢迎的数据。

# Text B

## Cloud Storage

扫码听课文

### 1. What is cloud storage

Cloud storage allows you to save data and files in an off-site location that you can access either through the public internet or a dedicated private network connection. To store the data that you transfer off-site becomes the responsibility of a third-party cloud provider. The provider hosts, secures, manages and maintains the servers and associated infrastructure and ensures you have access to the data whenever you need it.

Cloud storage delivers a cost-effective, scalable alternative to storing files on on-premise hard drives or storage networks. Computer hard drives can only store a finite amount of data. When users run out of storage, they need to transfer files to an external storage device. Traditionally, organizations built and maintained Storage Area Networks (SAN) to archive data and files. SAN are expensive to maintain because as stored data grows, companies have to invest in adding servers and infrastructure to accommodate the increased demand.

Cloud storage services provide elasticity, which means you can scale capacity as your data volumes increase or dial down capacity if necessary. By storing data in a cloud, organizations save by paying for storage technology and capacity as a service, rather than investing in the capital costs of building and maintaining in-house storage networks. You pay for only exactly the capacity you use. While your costs might increase over time to account for higher data volumes, you don't have to over provision storage networks in anticipation of increased data volume.

### 2. How does cloud storage work

Like on-premise storage networks, cloud storage uses servers to save data. However, the data is sent to servers at an off-site location. Most of the servers you use are virtual machines hosted on a physical server. As your storage needs increase, the provider creates new virtual servers to meet demand.

Typically, you connect to the storage cloud either through the internet or a dedicated private connection, using a web portal, website, or a mobile app. The server with which you connect forwards your data to a pool of servers located in one or more data centers, depending on the size of the cloud provider's operation.

As part of the service, providers typically store the same data on multiple machines for redundancy. This way, if a server is taken down for maintenance or suffers an outage, you can still access your data.

Cloud storage is available in private, public and hybrid clouds.

- Public cloud storage: In this model, you connect over the internet to a storage cloud that's maintained by a cloud provider and used by other companies. Providers typically make services accessible from just about any device, including smartphones and desktops and let you scale up and down as needed.

- Private cloud storage: Private cloud storage setups typically replicate the cloud model, but they reside within your network, leveraging a physical server to create instances of virtual servers to increase capacity. You can choose to take full control of an on-premise private cloud or engage a cloud storage provider to build a dedicated private cloud that you can access with a private connection. Organizations that might prefer private cloud storage include banks or retail companies due to the private nature of the data they process and store.

- Hybrid cloud storage: This model combines elements of private and public clouds, giving organizations a choice of which data to store in which cloud. For instance, highly regulated data subject to strict archiving and replication requirements is usually more suited to a private cloud environment, whereas less sensitive data (such as email that doesn't contain business secrets) can be stored in the public cloud. Some organizations use hybrid clouds to supplement their internal storage networks with public cloud storage.

### 3. Advantages and disadvantages of cloud storage

As with any other cloud-based technology, cloud storage offers some distinct advantages. But it also raises some concerns for companies, primarily over security and administrative control.

(1) Advantages of cloud storage

- Off-site management: Your cloud provider assumes responsibility for maintaining and protecting the stored data. This frees your staff from tasks associated with storage, such as procurement, installation, administration and maintenance. As such, your staff can focus on other priorities.

- Quick implementation: Using a cloud service accelerates the process of setting up and adding to your storage capabilities. With cloud storage, you can provision the service and start using it within hours or days, depending on how much capacity is involved.

- Cost-effective: As mentioned, you pay for the capacity you use. This allows your organization to treat cloud storage costs as an ongoing operating expense instead of a capital expense with the associated upfront investments and tax implications.

- Scalability: Growth constraints are one of the most severe limitations of on-premise storage. With cloud storage, you can scale up as much as you need. Capacity is virtually unlimited.

- Business continuity: Storing data offsite supports business continuity in the event that a natural disaster or an attack cuts access to your premises.

(2) Disadvantages of cloud storage

- Security: Security concerns are common with cloud-based services. Cloud storage providers try to secure their infrastructure with up-to-date technologies and practices, but occasional breaches have occurred, creating discomfort with users.

- Administrative control: Being able to view your data, access it and move it at will is another common concern with cloud resources. Offloading maintenance and management to a third party offers advantages but also can limit your control over your data.

- Latency: Delays in data transmission to and from the cloud can occur as a result of traffic congestion, especially when you use shared public internet connections. However, companies can minimize latency by increasing connection bandwidth.

- Regulatory compliance: Certain industries, such as healthcare and finance, have to comply with strict data privacy and archival regulations, which may prevent companies from using cloud storage for certain types of files, such as medical and investment records. If you can, choose a cloud storage provider that supports compliance with any industry regulations impacting your business.

## 4. Security

Cloud storage security is a serious concern, especially if your organization handles sensitive data like credit card information and medical records. You want assurances your data is protected from cyber threats with the most up-to-date methods available. You will want layered security solutions that include endpoint protection, content and email filtering and threat analysis, as well as best practices that comprise regular updates and patches. And you need well-defined access and authentication policies.

Most cloud storage providers offer baseline security measures that include access control, user authentication and data encryption. Ensuring these measures are in place is especially important when the data in question involves confidential business files, personnel records and intellectual property. Data subject to regulatory compliance may require added protection, so you need to check that your cloud storage provider of choice complies with all applicable regulations.

Whenever data travels, it is vulnerable to security risks. Your responsibility is that the data transferred to the cloud is safe. Companies can minimize risks by encrypting data in motion and using dedicated private connections (instead of the public internet) to connect with the cloud storage provider.

## 5. Backup

Data backup is as important as security. Businesses need to back up their data so they can access copies of files and applications and prevent interruptions to business if data is lost due to cyber attack, natural disaster, or human error.

Cloud-based data backup and recovery services have been popular from the early days of cloud-based solutions. Much like cloud storage itself, you access the service through the public internet or a private connection. Cloud backup and recovery services free organizations from the tasks involved in regularly replicating critical business data to make it readily available should you ever need it in the wake of data loss caused by a natural disaster, cyber attack or unintentional user error.

Cloud backup offers the same advantages to businesses as storage — cost-effectiveness, scalability and easy access. One of the most attractive features of cloud backup is automation. Asking users to continually backup their own data produces mixed results since some users always put it off or forget to do it. This creates a situation where data loss is inevitable. With automated backups, you can decide how often to back up your data, be it daily, hourly or whenever new data is introduced to your network.

Backing up data off-premise in a cloud offers an added advantage: distance. A building struck by a natural disaster, attack or some other calamity could lose its on-premise backup systems, making it impossible to recover lost data. Off-premise backup provides insurance against such an event.

## 📖 New Words

| | | |
|---|---|---|
| off-site | ['ɔːf saɪt] | adj.非现场的；站外的；异地的 |
| internet | ['ɪntənet] | n.互联网 |
| connection | [kə'nekʃn] | n.连接；联系 |
| transfer | [træns'fɜː] | v.传输，转移 |
| responsibility | [rɪ,spɒnsə'bɪləti] | n.责任；职责 |
| third-party | ['θɜːd'pɑːti] | adj.第三方的 |
| host | [həʊst] | n.主机 |
| | | v.托管 |
| deliver | [dɪ'lɪvə] | vt.交付；发表 |
| cost-effective | [,kɒst ɪ'fektɪv] | adj.有成本效益的，划算的 |
| alternative | [ɔːl'tɜːnətɪv] | adj.替代的；备选的 |
| | | n.可供选择的事物 |
| finite | ['faɪnaɪt] | adj.有限的 |
| elasticity | [,iːlæ'stɪsəti] | n.弹性；灵活性；伸缩性 |
| provision | [prə'vɪʒn] | n.预备，准备；供应 |
| virtual | ['vɜːtʃuəl] | adj.虚拟的；实际的 |
| demand | [dɪ'mɑːnd] | v.&n.需求，需要 |
| redundancy | [rɪ'dʌndənsi] | n.冗余；多余 |
| accessible | [ək'sesəbl] | adj.可访问的；易接近的 |
| replicate | ['replɪkeɪt] | vt.复制；重复 |
| capacity | [kə'pæsəti] | n.容积；生产量 |
| strict | [strɪkt] | adj.精确的；绝对的；严格的 |
| sensitive | ['sensətɪv] | adj.敏感的；易受影响的 |
| supplement | ['sʌplɪmənt] | vt.增强，补充 |
| internal | [ɪn'tɜːnl] | adj.内部的 |
| procurement | [prə'kjʊəmənt] | n.采购；获得，取得 |
| accelerate | [ək'seləreɪt] | v.加快，加速 |
| capital | ['kæpɪtl] | n.资本；资金 |
| constraint | [kən'streɪnt] | n.限制；约束 |
| unlimited | [ʌn'lɪmɪtɪd] | adj.无限的 |
| continuity | [,kɒntɪ'njuːəti] | n.连续性，连接 |

| disaster | [dɪ'zɑ:stə] | n.灾难 |
| occasional | [ə'keɪʒənl] | adj.偶尔的，临时的 |
| discomfort | [dɪs'kʌmfət] | n.不舒适，不舒服；不安 |
| latency | ['leɪtənsi] | n.延迟；潜伏 |
| congestion | [kən'dʒestʃən] | n.拥挤；阻塞 |
| bandwidth | ['bændwɪdθ] | n.带宽 |
| regulatory | ['regjələtəri] | adj.管理的；监管的 |
| compliance | [kəm'plaɪəns] | n.服从，听从；承诺 |
| comply | [kəm'plaɪ] | vi.遵从；依从，顺从 |
| archival | [ɑ:'kaɪvəl] | adj.档案的 |
| backup | ['bækʌp] | n.备份 |
| interruption | [ˌɪntə'rʌpʃn] | n.中断，打断 |
| recovery | [rɪ'kʌvəri] | n.恢复，复原 |
| unintentional | [ˌʌnɪn'tenʃənl] | adj.无意的，无心的 |
| inevitable | [ɪn'evɪtəbl] | adj.不可避免的，必然发生的 |
| strike | [straɪk] | v.&n.攻击 |
| calamity | [kə'læməti] | n.灾祸，灾难 |
| insurance | [ɪn'ʃʊərəns] | n.预防措施；保险，保险业；保险费 |

## Phrases

| cloud storage | 云存储 |
| private network | 私有网络，专用网络 |
| cloud provider | 云提供商 |
| hard drive | 硬盘驱动器 |
| external storage device | 外部存储设备 |
| in anticipation of | 期待着……；预计到…… |
| hybrid cloud | 混合云 |
| scale up | 按比例增加，按比例提高 |
| public cloud | 公共云 |
| private cloud | 私有云，专用云 |
| focus on | 致力于；使聚焦于 |
| cyber attack | 网络攻击 |

## Abbreviations

| SAN (Storage Area Network) | 存储区域网络 |

# Reference Translation

# 云 存 储

## 1. 什么是云存储

云存储允许你可以将数据和文件在异地保存，你可以通过公共互联网或专用的私有网络连接进行访问。第三方云提供商负责把存储在异地的数据传输给你。提供者负责托管、保护、管理和维护服务器及相关基础结构，并确保你可以随时访问数据。

云存储提供了一种经济高效、可扩展的替代方案，可以将文件存储在本地硬盘驱动器或者存储在网络上。计算机硬盘驱动器只能存储有限数量的数据。当用户的存储空间用完时，他们需要将文件传输到外部存储设备。传统上，组织会建立和维护存储区域网络（SAN）来存档数据和文件。SAN 的维护成本很高，因为随着存储数据的增长，公司必须增加服务器和基础架构的投资来适应不断增长的需求。

云存储服务提供了弹性，这意味着你可以在数据量增加的同时扩展容量，或者在必要时降低容量。通过将数据存储在云中，组织可以通过为存储技术和容量即服务付费来节省资金，而不是投资建设和维护内部存储网络。你只需为使用的容量付费。尽管随着时间的流逝，为了容纳更大的数据量，成本可能会增加，但不必为预计增加的数据量超前投资。

## 2. 云存储如何运作

与本地存储网络一样，云存储使用服务器来保存数据。但是，数据将发送到异地服务器。你使用的大多数服务器是托管在物理服务器上的虚拟机。随着存储需求的增加，提供商将创建新的虚拟服务器来满足需求。

通常，你可以使用网络门户、网站或移动应用程序，通过互联网或专用私有网连接到存储云。你所连接的服务器会将你的数据转发到位于一个或多个数据中心的服务器池中，具体取决于云提供商的运营规模。

作为服务的一部分，提供商通常将相同的数据存储在多台计算机上以实现冗余。这样，如果服务器因维护或发生故障而关闭时，你仍然可以访问数据。

云存储可在私有云、公共云和混合云中使用。

- 公共云存储：在这种模式下，你可以通过互联网连接到由云提供商维护并由其他公司使用的存储云。提供商通常提供可以从几乎任何设备（包括智能手机和台式机）访问的服务，并可以根据需要进行扩展和缩减。
- 私有云存储：私有云存储设置通常复制云模型，但它们驻留在你的网络中，这样就可以利用物理服务器创建虚拟服务器实例以增加容量。你可以选择完全控制本地私有云，也可以聘用云存储提供商来构建可以通过私有连接访问的专用私有云。由于处理和存储的数据具有私有性，因此更喜欢私有云存储的组织可能包括银行或零售公司。
- 混合云存储：此模型结合了私有云和公共云的元素，使组织可以选择将哪些数据存储在哪个云中。例如，对归档和复制有严格控制要求的数据通常更适合私有云环境，而敏感度较低的数据（如不包含商业机密的电子邮件）可以存储在公共云中。一些组织使用混合云，用公共云存储来补充其内部存储网络。

**3. 云存储的优点和缺点**

与任何其他基于云的技术一样，云存储具有一些明显的优势。但这也给公司带来了一些担忧，主要是在安全性和管理控制方面。

（1）云存储的优点

- 异地管理：你的云提供商负责维护和保护存储的数据。这使你的员工摆脱了与存储相关的任务，例如采购、安装、管理和维护。因此，你的员工可以专注于其他优先事项。
- 快速实施：使用云服务可以加快设置和添加存储功能的过程。借助云存储，你可以根据所涉及的容量在几小时或几天内配置服务并开始使用它。
- 经济高效：如上所述，你需要为使用的容量付费。这使你的组织可以将云存储成本视为持续的运营支出，而不是将资本支出与相关的前期投资和税费相关联。
- 可扩展性：增长限制是本地存储的最严重限制之一。借助云存储，你可以根据需要进行扩展。容量实际上是无限的。
- 业务连续性：遇到自然灾害或因为攻击而不能访问你的场所时，存储在异地的数据可以支持业务连续性。

（2）云存储的缺点

- 安全性：安全问题在基于云的服务中很常见。云存储提供商尝试使用最新的技术和实践来保护其基础架构，但是偶尔会发生违规行为，给用户带来不便。
- 管理控制：能够查看、访问和随意移动数据是云资源的另一个常见问题。将维护和管理工作转移给第三方可以带来好处，但同时也可能限制你对数据的控制。
- 延迟：由于流量拥塞，尤其是在使用共享的公共互联网连接时，往返云的数据传输可能会发生延迟。但是，公司可以通过增加连接带宽来最大程度减少延迟。
- 合规性：某些行业（如医疗保健和金融业）必须遵守严格的数据隐私和档案法规，这可能会阻止公司将云存储用于某些类型的文件（如医疗和投资记录）。如果可以，请选择支持符合你业务的各种行业法规的云存储提供商。

**4. 安全性**

云存储安全是一个严重的问题，当你的组织处理诸如信用卡信息和病历之类的敏感数据时尤其重要。你需要使用最新的方法来确保数据免受网络威胁。你需要分层的安全解决方案，其中包括端点保护、内容和电子邮件过滤以及威胁分析和定期更新与打补丁。你需要定义明确的访问和身份验证策略。

大多数云存储提供商都提供基准安全措施，包括访问控制、用户身份验证和数据加密。当相关数据涉及机密业务档案、人员记录和知识产权时，确保采取适当措施尤其重要。符合法规要求的数据可能需要进一步的保护，因此你需要核实所选的提供商是否符合所有适用法规。

每当数据传输时，它都容易受到安全风险的影响。你的责任是确保传输到云的数据是安全的。公司可以通过加密动态数据并使用专用的私有连接（而不是公共互联网）与云存储提供商进行连接，从而将风险降到最低。

**5. 备份**

数据备份与安全同等重要。企业需要备份其数据，以便它们可以访问文件和应用程序的副本，并防止由于网络攻击、自然灾害或人为错误而导致数据丢失情况下的业务中断。

从早期的基于云的解决方案开始，基于云的数据备份和恢复服务就很受欢迎。就像云存储本身一样，你可以通过公共互联网或专用连接访问该服务。云备份和恢复服务将组织从定期复制关键业务数据的任务中解放出来，以便在因自然灾害、网络攻击或意外用户错误而导致数据丢失后随时使用。

云备份为企业提供的优势与存储相同——经济高效、可扩展性和易于访问。云备份最吸引人的特点之一就是自动化。由于某些用户总是拖延或忘记这样做，因此要求用户不断备份自己的数据，会产生不同的结果。这造成了数据丢失的不可避免。使用自动备份，你可以决定备份数据的频率，是每天、每小时或每当将新数据引入网络时进行备份。

在云端异地备份数据具有一个额外的优势：距离。受自然灾害、攻击或其他灾难袭击的建筑物可能会丢失其内部备份系统，从而无法恢复丢失的数据。异地备份为此类事件提供了保障。

# Exercises

**[Ex. 1]** 根据 Text A 填空。

1. When determining what information you want to collect, the first thing you need to do is to choose _____ . You'll need to decide _____, who you want to collect it from and _____.

2. If you're tracking data for a specific campaign, however, you'll track it _____. In these instances, you should have _____ for when you'll start and _____.

3. To select the right collection method, you'll need to consider _____you want to collect, _____ over which you'll obtain it and _____you are determined.

4. Once you have finalized your plan, you can _____ and _____. You can _____ your data in your DMP (Data Management Platform).

5. A survey consists of _____ respondents can answer _____ and often gives participants a list of responses to choose from. You can conduct surveys _____, _____, _____ or in person.

6. Your business website and your app if you have one, are excellent tools for _____. When someone visits your website, they create _____.

7. You may store transactional records in _____. This information can give you insights about _____, _____, how often people typically purchase from you and more.

8. The software you use to place your ads will likely give you data about _____, _____, _____ and more.

9. You can also monitor mentions of your brand on social media by _____, _____ or _____.

10. The most straightforward way to do in-store traffic monitoring is _____ on the door to help you keep track of _____ throughout the day.

**[Ex. 2]** 根据 **Text B** 回答以下问题。

1. What does cloud storage allow you to do?

2. Why are SAN expensive?

3. How can you connect to the storage cloud typically?

4. In public cloud storage, what do you connect to over the internet?

5. What do organizations that might prefer private cloud storage include?

6. What kind of data is suited to a private cloud environment? What kind of data can be stored in the public cloud?

7.What are the advantages of cloud storage?

8.What are the disadvantages of cloud storage?

9. What do most cloud storage providers offer? When is ensuring these measures are in place especially important?

10. Why do businesses need to back up their data?

**[Ex. 3]** 词汇英译中

| | |
|---|---|
| 1. data collection | 1. _____ |
| 2. collect | 2. _____ |
| 3. cloud storage | 3. _____ |
| 4. demonstrate | 4. _____ |
| 5. cyber attack | 5. _____ |
| 6. establish | 6. _____ |
| 7. hybrid cloud | 7. _____ |
| 8. link | 8. _____ |
| 9. private cloud | 9. _____ |
| 10. monitoring | 10. _____ |

**[Ex. 4] 词汇中译英**

1. 公共云　　　　　　　　　　　　1. _____
2. 在线的；联网的；联机的　　　　　2. _____
3. 云提供商　　　　　　　　　　　　3. _____
4. 网站　　　　　　　　　　　　　　4. _____
5. 数据点　　　　　　　　　　　　　5. _____
6. 备份　　　　　　　　　　　　　　6. _____
7. 建立，设立；安排　　　　　　　　7. _____
8. 带宽　　　　　　　　　　　　　　8. _____
9. 私有网络，专用网络　　　　　　　9. _____
10. 容积；生产量　　　　　　　　　　10. _____

**[Ex. 5] 短文翻译**

### 5 Steps to Collect Big Data

Today, many companies collect big data to analyze and interpret daily transactions and traffic data, aiming to keep track of the operations, forecast needs or implement new programs. But how to collect big data directly?

There may be a lot of data collection methods and you may feel quite confused. Here I will introduce the general steps to collect big data.

Step 1: Gather data.

There are many ways to gather data according to different purposes. For example, you can buy data from Data-as-Service companies or use a data collection tool to gather data from websites.

Step 2: Store data.

After gathering the big data, you can put the data into databases or storage services for further processing. Usually, this step requires investment in the physical foundation as well as cloud services. Some data collection tools provide unlimited cloud storage after data is gathered, which greatly saves local resources and makes data easy to access from anywhere.

Step 3: Clean up data.

Since there may be noisy information you don't need, you need to pick up the one that meets your needs. This step is to sort the data, including cleaning up, concatenating and merging the data.

Step 4: Reorganize data.

You need to reorganize the data after cleaning it up for further use. Usually, you need to turn the unstructured or semi-unstructured formats into structured formats like Hadoop and HDFS.

Step 5: Verify data.

To make sure the data you get is right and meaningful, you need to verify the data. Choose some samples to see whether it works.

These are the general steps to collect big data. However, to collect the data, analyze it and glean insights into markets is not as easy as it seems. Data collection tools like Octoparse help make this process so much easier. They allow users to gather clean and structured data automatically so there is no need to clean it up or reorganize it. After the data is collected, it can be stored in cloud databases, which can be accessed anytime from anywhere.

# Reading Material

## Data Center

A data center is a facility composed of[1] networked computers and storage that businesses and other organizations use to organize, process, store and disseminate[2] large amounts of data. A business typically relies heavily upon the applications, services and data contained within a data center, making it a focal point and critical asset for everyday operations.

### 1. How data centers work

Data centers are not a single thing, but rather a conglomeration[3] of elements. At a minimum, data centers serve as the principal repositories[4] for all manner of IT equipment, including servers, storage subsystems, networking switches[5], routers[6] and firewalls, as well as the cabling and physical racks[7] used to organize and interconnect the IT equipment. A data center must also contain an adequate[8] infrastructure, such as power distribution and supplemental power subsystems. This also includes electrical switching, uninterruptable power supplies[9], backup generators[10], ventilation[11] and data center cooling systems, such as in-row cooling configurations and computer room air conditioners; and adequate provisioning for network carrier connectivity. All of this demands a physical facility with physical security and sufficient square footage to house the entire collection of infrastructure and equipment.

### 2. What is data center consolidation[12]

There is no requirement for a single data center, and modern businesses may use two or more data center installations across multiple locations for greater resilience and better application performance, which lowers latency by locating workloads closer to users.

Conversely, a business with multiple data centers may opt to consolidate data centers, reducing the number of locations in order to minimize the costs of IT operations. Consolidation typically

---

1　be composed of: 由……组成
2　disseminate [dɪ'semɪneɪt] vt.分发
3　conglomeration [kən,glɒmə'reɪʃn] n.聚集
4　repository [rɪ'pɒzətri] n.仓库
5　switch [swɪtʃ] n.交换机
6　router ['ru:tə] n.路由器
7　rack [ræk] n.机架，支架
8　adequate ['ædɪkwət] adj.足够的；合格的；合乎需要的
9　uninterruptable power supply: 不间断电源
10　generator ['dʒenəreɪtə] n.发电机
11　ventilation [,ventɪ'leɪʃən] n.通风设备
12　consolidation [kən,sɒlɪ'deɪʃən] n.整合，合并

occurs during mergers and acquisitions when the majority business doesn't need the data centers owned by the subordinate business.

### 3. What is data center colocation

Data center operators can also pay a fee to rent server space in a colocation facility. Colocation[13] is an appealing option for organizations that want to avoid the large capital expenditures associated with building and maintaining their own data centers. Today, colocation providers are expanding their offerings to include managed services, such as interconnectivity, allowing customers to connect to the public cloud.

### 4. Data center tiers

Data centers are not defined by their physical size or style. Small businesses may operate successfully with several servers and storage arrays networked within a convenient closet or small room, while major computing organizations, such as Facebook, Amazon or Google, may fill an enormous warehouse space with data center equipment and infrastructure. In other cases, data centers can be assembled in mobile installations, such as shipping containers[14], also known as data centers in a box, which can be moved and deployed as required.

However, data centers can be defined by various levels of reliability or resilience, sometimes referred to as data center tiers. In 2005, the American National Standards Institute (ANSI) and the Telecommunications Industry Association (TIA) published standard ANSI/TIA-942 "Telecommunications Infrastructure Standard for Data Centers", which defined four tiers of data center design and implementation guidelines. Each subsequent tier is intended to provide more resilience, security and reliability than the previous tier. For example, a tier 1 data center is little more than a server room, while a tier 4 data center offers redundant[15] subsystems and high security.

### 5. Data center architecture and design

Although almost any suitable space could conceivably serve as a "data center", the deliberate design and implementation of a data center requires careful consideration. Beyond the basic issues of cost and taxes, sites are selected based on a multitude of criteria, such as geographic location, seismic and meteorological[16] stability, access to roads and airports, availability of energy and telecommunications.

Once a site is secured, the data center architecture can be designed with attention to the mechanical and electrical infrastructure as well as the composition and layout of the IT equipment. All of these issues are guided by the availability and efficiency goals of the desired data center tier.

### 6. Energy consumption[17] and efficiency

Data center designs also recognize the importance of energy efficiency. A simple data center

---

13    colocation [ˌkələʊ'keɪʃən] *n*.托管；场地出租

14    container [kən'teɪnə] *n*.容器；集装箱

15    redundant [rɪ'dʌndənt] *adj*.冗余的

16    meteorological [ˌmiːtɪərə'lɒdʒɪkl] *adj*.气象的

17    consumption [kən'sʌmpʃn] *n*.消耗

may need only a few kilowatts of energy, but an enterprise-scale data center installation can demand tens of megawatts or more. Today, the green data center, which is designed for minimum environmental impact through the use of low-emission building materials, catalytic converters and alternative energy technologies, is growing in popularity.

## 7. Data center security and safety

Data center designs must also implement sound safety and security practices. For example, safety is often reflected in the layout of doorways and access corridors[18], which must accommodate the movement of large, unwieldy IT equipment as well as permit employees to access and repair the infrastructure.

Fire suppression is another key safety area, and the extensive use of sensitive, high-energy electrical and electronic equipment precludes common sprinklers[19]. Instead, data centers often use environmentally friendly chemical fire suppression systems[20], which effectively starve a fire of oxygen while mitigating collateral damage to the equipment. Because the data center is also a core business asset, comprehensive security measures, like badge access and video surveillance[21], help to detect and prevent malfeasance by employees, contractors and intruders.

## 8. Data center infrastructure management and monitoring

Modern data centers make extensive[22] use of monitoring and management software. Software including data center infrastructure management tools allow remote IT administrators to oversee the facility and equipment, measure performance, detect failures and implement a wide array of corrective actions without ever physically entering the data center room.

The growth of virtualization[23] has added another important dimension to data center infrastructure management. Virtualization now supports the abstraction[24] of servers, networks and storage, allowing every computing resource to be organized into pools without regard to their physical location. Administrators can then provision workloads, storage instances and even network configuration from those common resource pools. When administrators no longer need those resources, they can return them to the pool for reuse. All of the actions network, storage and server virtualization accomplish can be implemented through software, giving traction[25] to the term software-defined data center.

## 9. Data center vs. cloud

Data centers are increasingly implementing private cloud software, which builds on virtualization to add a level of automation, user self-service and billing/chargeback[26] to data center

---

18  corridor ['kɒridɔː] n.走廊，通道

19  sprinkler ['sprɪŋklə] n.洒水器；（建筑物内的）自动喷水灭火装置

20  chemical fire suppression system: 化学灭火系统

21  surveillance [sɜː'veɪləns] n.监督，监视

22  extensive [ɪk'stensɪv] adj.广阔的；广泛的；大量的

23  virtualization [vɜːtʃʊəlaɪ'zeɪʃn] n.虚拟化

24  abstraction [æb'strækʃn] n.抽象

25  traction ['trækʃn] n.引出；牵引

26  chargeback [t'ʃɑːdʒbæk] n.退款；拒付

administration. The goal is to allow individual users to provision workloads and other computing resources on demand without IT administrative intervention.

It is also increasingly possible for data centers to interface with public cloud providers. Platforms such as Microsoft Azure emphasize[27] the hybrid use of local data centers with Azure or other public cloud resources. The result is not an elimination of data centers, but rather the creation of a dynamic environment that allows organizations to run workloads locally or in the cloud or to move those instances to or from the cloud as desired.

---

27    emphasize ['emfəsaɪz] v.强调；重视

# Unit 4

## Text A

### Database

扫码听课文

A database is an organized collection of structured information, or data, typically stored electronically in a computer system. A database is usually controlled by a database management system (DBMS). Together, the data and the DBMS, along with the applications that are associated with them, are referred to as a database system, often shortened to just database.

Data within the most common databases today is typically modeled in rows and columns to make processing and data querying efficient. The data can then be easily accessed, managed, modified, updated, controlled and organized. Most databases use Structured Query Language (SQL) for writing and querying data.

### 1. What is structured query language (SQL)

SQL is a specialized programming language which is standardized to be used for managing relational databases and performing various operations on the data. There are various uses of SQL which includes modifying database table and index structures; adding, updating and deleting rows of data; and retrieving various subsets of information from a database for transaction processing and analytics applications. There are specialized queries and operations which operates in the form of commands and commonly known as SQL statements like select, add, insert, update, delete, create, alter and truncate.

SQL is also a domain-specific language used for programming and designing in data held in a Relational DataBase Management System (RDBMS). It is particularly useful in handling structured data where there are relations between different entities/variables of the data.

SQL is originally based upon relational algebra and tuple relational calculus. It consists of many types of statements, commonly known as a Data Query Language (DQL), a Data Definition Language (DDL), a Data Control Language (DCL) and a Data Manipulation Language (DML). The scope of SQL includes data query, data manipulation (insert, update and delete), data definition (schema creation and modification) and data access control. Also, SQL is described as a declarative Language (4GL), because it also includes procedural elements.

### 2. Types of databases

There are many different types of databases. The best database for a specific organization depends on how the organization intends to use the data.

- Relational databases. Relational databases became dominant in the 1980s. Items in a relational database are organized as a set of tables with columns and rows. Relational database technology provides the most efficient and flexible way to access structured information.

- Object-oriented databases. Information in an object-oriented database is represented in the form of objects, as in object-oriented programming.
- Distributed databases. A distributed database consists of two or more files located in different sites. The database may be stored on multiple computers located in the same physical location or scattered over different networks.
- Data warehouses. A central repository for data, a data warehouse is a type of database specifically designed for fast query and analysis.
- NoSQL databases. A NoSQL or nonrelational database, allows unstructured and semistructured data to be stored and manipulated (in contrast to a relational database, which defines how all data inserted into the database must be composed). NoSQL databases grew popular as web applications became more common and more complex.
- Graph databases. A graph database stores data in terms of entities and the relationships between entities.
- OLTP databases. An OLTP database is a speedy, analytic database designed for large numbers of transactions performed by multiple users.

These are only a few of the several dozen types of databases in use today. Some less common databases are tailored to very specific scientific, financial or other functions. In addition to the different database types, changes in technology development approaches and dramatic advances such as the cloud and automation are propelling databases in entirely new directions. Some of the latest databases are as follows.

- Open source database. An open source database system is one whose source code is open source, such databases could be SQL or NoSQL databases.
- Cloud database. A cloud database is a collection of data, either structured or unstructured, that resides on a private, public or hybrid cloud computing platform. There are two types of cloud database models: traditional and DataBase as a Service (DBaaS). With DBaaS, administrative tasks and maintenance are performed by a service provider.
- Multimodel database. Multimodel database combines different types of database models into a single, integrated back end. This means it can accommodate various data types.
- Document/JSON database. Designed for storing, retrieving and managing document-oriented information, document databases are a modern way to store data in JSON format rather than rows and columns.
- Self-driving database. Self-driving databases (also known as autonomous databases) are the newest and most groundbreaking type of database that are cloud-based and use machine learning to automate database tuning, security, backups, updates and other routine management tasks traditionally performed by database administrators.

### 3. What is a database management system

DBMS software primarily functions as an interface between the end user and the database, simultaneously managing the data, the database engine and the database schema in order to facilitate the organization and management of data.

Though functions of DBMS vary greatly, the general features and capabilities of the DBMS should include a user accessible catalog describing metadata, DBMS library management system, data abstraction and independence, data security, logging and auditing of activity, support for concurrency and transactions, support for authorization of access, access support from remote locations, DBMS data recovery support in the event of damage and enforcement of constraints to ensure the data follows certain rules.

A database management system functions through the use of system commands, first receiving instructions from a database administrator in DBMS, then instructing the system accordingly, either to retrieve data, modify data or load existing data from the system. Popular DBMS examples include cloud-based database management systems, In-Memory DataBase Management Systems (IMDBMS), Columnar DataBase Management Systems (CDBMS) and NoSQL in DBMS.

### 4. What is a MySQL database

MySQL is an open source relational database management system based on SQL. It was designed and optimized for web applications and can run on any platform. As new and different requirements emerged with the internet, MySQL became the platform of choice for web developers and web-based applications. Because it's designed to process millions of queries and thousands of transactions, MySQL is a popular choice for e-commerce businesses that need to manage multiple money transfers. On-demand flexibility is the primary feature of MySQL.

### 5. Database challenges

Today's large enterprise databases often support very complex queries and are expected to deliver nearly instant responses to those queries. As a result, database administrators are constantly called upon to employ a wide variety of methods to help improve performance. Some common challenges that they face are as follows.

- Absorbing significant increases in data volume. The explosion of data coming in from sensors, connected machines and dozens of other sources makes it difficult for database administrators to effectively manage and organize their companies' data.
- Ensuring data security. Data breaches are happening everywhere these days, and hackers are getting more inventive. It's more important than ever to ensure that data is secure but also easily accessible to users.
- Keeping up with demand. In today's fast-moving business environment, companies need real-time access to their data to support timely decision-making and to take advantage of new opportunities.
- Managing and maintaining the database and infrastructure. Database administrators must continually watch the database for problems and perform preventative maintenance, as well as apply software upgrades and patches. As databases become more complex and data

volumes grow, companies are faced with the expense of hiring additional talents to monitor and tune their databases.

● Removing limits on scalability. A business needs to grow if it's going to survive, and its data management must grow along with it. But it's very difficult for database administrators to predict how much capacity the company will need, particularly with on-premises databases.

Addressing these challenges can be time-consuming and can prevent database administrators from performing more strategic functions.

## New Words

| database | ['deɪtəbeɪs] | n.数据库 |
| electronically | [ɪˌlek'trɒnɪkli] | adv.电子地 |
| control | [kən'trəʊl] | vt.控制；管理 |
| modify | ['mɒdɪfaɪ] | v.修改，改变 |
| delete | [dɪ'li:t] | v.删除 |
| subset | ['sʌbset] | n.子集 |
| command | [kə'mɑ:nd] | n.命令 |
| statement | ['steɪtmənt] | n.语句；声明 |
| select | [sɪ'lekt] | vt.选择 |
| insert | [ɪn'sɜ:t] | vt.插入；嵌入 |
| create | [kri'eɪt] | vt.建立 |
| alter | ['ɔ:ltə] | vt.改变，更改 |
| truncate | [trʌŋ'keɪt] | vt.截断，缩短 |
| variable | ['veəriəbl] | n.变量 |
| | | adj.变量的；变化的，可变的 |
| tuple | [tʌpl] | n.元组，数组 |
| calculus | ['kælkjələs] | n.运算，计算 |
| manipulation | [məˌnɪpjʊ'leɪʃn] | n.操作；控制 |
| declarative | [dɪ'klærətɪv] | adj.声明的，陈述的 |
| procedural | [prə'si:dʒərəl] | adj.程序的，过程的 |
| dominant | ['dɒmɪnənt] | adj.占优势的；统治的，支配的 |
| object-oriented | ['ɒbdʒɪkt 'ɔ:rɪəntɪd] | adj.面向对象的 |
| distribute | [dɪ'strɪbju:t] | vt.分布，分配；散发，分发 |
| scatter | ['skætə] | v.散开，分散 |
| repository | [rɪ'pɒzətri] | n.仓库；储藏室 |
| dramatic | [drə'mætɪk] | adj.戏剧性的；引人注目的；突然的 |
| propel | [prə'pel] | vt.推进，推动，驱动 |

| administrative | [əd'mɪnɪstrətɪv] | *adj.*管理的，行政的 |
| multimodel | ['mʌltɪ'mɒdl] | *n.*多模式 |
| groundbreaking | ['graʊndbreɪkɪŋ] | *adj.*开创性的，突破性的 |
| routine | [ru:'ti:n] | *n.*常规；例行程序 |
| | | *adj.*例行的；常规的 |
| interface | ['ɪntəfeɪs] | *n.*界面；接口 |
| simultaneously | [ˌsɪməl'teɪnɪəsli] | *adv.*同时地 |
| engine | ['endʒɪn] | *n.*引擎，发动机 |
| facilitate | [fə'sɪlɪteɪt] | *vt.*促进，助长；使容易 |
| abstraction | [æb'strækʃn] | *n.*抽象；抽象化 |
| independence | [ˌɪndɪ'pendəns] | *n.*独立，自主 |
| concurrency | [kən'kʌrənsi] | *n.*并发（性） |
| authorization | [ˌɔ:θəraɪ'zeɪʃn] | *n.*授权，批准 |
| remote | [rɪ'məʊt] | *adj.*远程的，遥远的 |
| enforcement | [ɪn'fɔ:smənt] | *n.*强制，实施，执行 |
| instruction | [ɪn'strʌkʃn] | *n.*指令 |
| optimize | ['ɒptɪmaɪz] | *vt.*使最优化 |
| developer | [dɪ'veləpə] | *n.*开发者 |
| e-commerce | [i:'kɒmɜ:s] | *n.*电子商务 |
| instant | ['ɪnstənt] | *n.*瞬间，顷刻 |
| | | *adj.*立即的 |
| hacker | ['hækə] | *n.*黑客 |
| inventive | [ɪn'ventɪv] | *adj.*发明的；有创造力的 |
| preventative | [prɪ'ventətɪv] | *adj.*预防性的 |
| talent | ['tælənt] | *n.*人才；天才 |
| remove | [rɪ'mu:v] | *v.*消除，移除 |
| time-consuming | ['taɪm kən'sju:mɪŋ] | *adj.*费时的，旷日持久的 |

## ✎ Phrases

| computer system | 计算机系统 |
| be associated with | 与……相关联 |
| a series of | 一系列；一连串 |
| programming language | 程序设计语言，编程语言 |
| data definition | 数据定义 |
| data access control | 数据存取控制，数据访问控制 |
| be described as | 被描述为 |
| declarative language | 说明性语言；声明性语言 |
| intend to | 打算，期望 |

| | |
|---|---|
| be organized as | 被组织成 |
| object-oriented database | 面向对象数据库 |
| object-oriented programming | 面向对象程序设计 |
| distributed database | 分布式数据库 |
| data warehouse | 数据仓库 |
| nonrelational database | 非关系数据库 |
| graph database | 图形数据库 |
| multiple user | 多用户 |
| open source | 开源 |
| source code | 源代码 |
| cloud database | 云数据库 |
| service provider | 服务提供商 |
| multimodel database | 多模式数据库 |
| back end | 后端 |
| document database | 文档数据库 |
| self-driving database | 自动驾驶数据库 |
| autonomous database | 自治数据库 |
| machine learning | 机器学习 |
| end user | 最终用户，终端用户 |
| cloud-based database management system | 基于云的数据库管理系统 |
| keep up with | 跟上 |

## ✎ Abbreviations

| | |
|---|---|
| DQL (Data Query Language) | 数据查询语言 |
| DDL (Data Definition Language ) | 数据定义语言 |
| DML (Data Manipulation Language) | 数据操作语言 |
| 4GL (Fourth Generation Language) | 第四代语言 |
| DBaaS (DataBase as a Service) | 数据库即服务 |
| JSON (JavaScript Object Notation) | JS 对象表示法 |
| IMDBMS (In-Memory DataBase Management System) | 内存数据库管理系统 |
| CDBMS (Columnar DataBase Management System) | 列式数据库管理系统 |

## Reference Translation

# 数 据 库

　　数据库是结构化信息或数据的有组织的集合，它通常以电子方式存储在计算机系统中。数据库通常由数据库管理系统（DBMS）管理。数据和 DBMS 以及与之关联的应用程序一起被称为数据库系统，经常简称为数据库。

　　当前，最常见的数据库中的数据通常以行和列建模，用来有效处理和查询数据。然后可以轻松地访问、管理、修改、更新、控制和组织数据。大多数的数据库使用结构化查询语言

（SQL）来编写和查询数据。

## 1. 什么是结构化查询语言（SQL）

SQL（结构化查询语言）是一种专用的编程语言，已标准化，可用于管理关系数据库和对数据执行各种操作。SQL 有多种用途，包括修改数据库表和索引结构。添加、更新和删除数据行；从数据库中检索各种信息子集，以用于事务处理和分析应用程序。有一些专门的查询和操作，它们以命令的形式进行操作，通常称为 SQL 语句，例如选择、添加、插入、更新、删除、创建、更改和截断。

SQL 还是一种特定区域的语言，用于对关系数据库管理系统（RDBMS）中保存的数据进行编程和设计。它在处理结构化数据时特别有用，其中数据的不同实体/变量之间存在关系。

SQL 最初基于关系代数和元组关系运算，它由多种类型的语句组成，通常称为数据查询语言（DQL）、数据定义语言（DDL）、数据控制语言（DCL）和数据操作语言（DML）。SQL 的范围包括数据查询、数据操作（插入、更新和删除）、数据定义（模式创建和修改）以及数据访问控制。SQL 也被描述为一种声明性语言（4GL），因为它还包含过程元素。

## 2. 数据库类型

有许多不同类型的数据库。对于特定组织，哪种数据库最好，这取决于组织打算如何使用数据。

- 关系数据库。关系数据库在 20 世纪 80 年代占主导地位。关系数据库中的项目被组织为一组具有列和行的表。关系数据库技术提供了访问结构化信息的最有效和最灵活的方法。
- 面向对象的数据库。同面向对象的编程一样，面向对象的数据库中的信息以对象的形式表示。
- 分布式数据库。分布式数据库由位于不同站点中的两个或多个文件组成。该数据库可以存储在位于相同物理位置的多台计算机上，也可以分散存储在不同的网络上。
- 数据仓库。数据仓库是数据的中央存储库，是专门为快速查询和分析而设计的一种数据库。
- NoSQL 数据库。NoSQL 或非关系数据库允许存储和处理非结构化与半结构化数据（与关系数据库相反，关系数据库定义了插入数据库的所有数据的必须组成方式）。随着网络应用程序越来越普遍和复杂，NoSQL 数据库越来越流行。
- 图形数据库。图形数据库根据实体以及实体之间的关系存储数据。
- OLTP 数据库。OLTP 数据库是一种快速的分析数据库，设计用于由多个用户执行的大量事务。

这些只是当今使用的几十种数据库中的几种。一些不太常见的数据库则针对非常具体的科学、财务或其他功能进行了定制。除了不同的数据库类型之外，技术开发方法的变化以及诸如云和自动化之类的显著进步正在推动数据库朝着全新的方向发展。一些最新的数据库如下。

- 开源数据库。开源数据库系统是指其源代码为开源的系统，这样的数据库可以是 SQL 或 NoSQL 数据库。

- 云数据库。云数据库是结构化或非结构化数据的集合。它驻留于私有、公共或混合云计算平台上。云数据库模型有两种类型：传统模型和数据库即服务（DBaaS）模型。使用 DBaaS，管理任务和维护由服务提供商执行。
- 多模型数据库。多模型数据库将不同类型的数据库模型组合到单个集成的后端中。这意味着它可以容纳各种数据类型。
- 文档/ JSON 数据库。专为存储、检索和管理面向文档的信息而设计，文档数据库是一种以 JSON 格式而不是行和列存储数据的现代方法。
- 自动驾驶数据库。自动驾驶数据库（也称为自治数据库）是最新、最有突破性的数据库类型，它基于云，并使用机器学习来自动执行数据库调整、安全性、备份、更新和其他传统上由数据库管理员执行的日常管理任务。

**3. 什么是数据库管理系统**

DBMS 软件的主要功能是充当最终用户和数据库之间的接口，同时管理数据、数据库引擎和数据库模式，以帮助组织和管理数据。

尽管 DBMS 的功能差异很大，但 DBMS 的通用功能应包括描述元数据的用户可访问目录、DBMS 库管理系统、数据抽象和独立性、数据安全性、活动的日志记录和审计、支持并发和事务、支持访问授权、支持远程访问、数据损坏时支持 DBMS 恢复数据以及强制执行约束，以确保数据遵循某些规则。

数据库管理系统通过使用系统命令来运行，首先从 DBMS 的数据库管理员那里接收指令，然后相应地指令系统从数据库中检索数据、修改数据或加载现有数据。流行的 DBMS 示例包括基于云的数据库管理系统、内存数据库管理系统（IMDBMS）、列式数据库管理系统（CDBMS）和 DBMS 中的 NoSQL。

**4. 什么是 MySQL 数据库**

MySQL 是基于 SQL 的开源关系数据库管理系统。它是为网络应用程序而设计和优化的，可以在任何平台上运行。随着互联网出现了新的不同需求，MySQL 成为网络开发人员和基于网络应用程序的首选平台。由于 MySQL 是为处理数百万查询和数千事务而设计的，因此它是需要管理多个转账的电子商务企业的热门选择。按需灵活性是 MySQL 的主要功能。

**5. 数据库挑战**

当今的大型企业数据库通常支持非常复杂的查询，并且有望对这些查询提供几乎即时的响应。因此，不断地要求数据库管理员采用各种各样的方法来帮助提高性能。他们面临的一些常见挑战如下。

- 吸收的数据量急剧增加。来自传感器、连接的机器以及许多其他来源的数据激增，使数据库管理员有效管理和组织公司的数据很困难。
- 确保数据安全。如今，数据泄露无处不在，黑客正变得越来越有创造力。确保数据安全并让用户轻松访问比以往任何时候都更为重要。
- 跟上需求。在当今瞬息万变的业务环境中，公司需要实时访问其数据以支持及时的决策并利用新机遇。
- 管理和维护数据库与基础架构。数据库管理员必须持续观察数据库、发现其中的问题并执行预防性维护，也要对应用软件升级和给程序打补丁。随着数据库变得越来越复杂且数据量不断增长，公司面临聘请更多人才来监管和调整数据库的开销。
- 消除对扩展性的限制。企业要生存就必须发展，其数据管理也必须随之发展。但是，数据库管理员很难预测公司将需要多少容量，尤其是对于本地数据库而言。

解决这些挑战可能很耗时，并且可能阻止数据库管理员执行更具战略意义的任务。

# Text B

## Data Warehouse

扫码听课文

### 1. What is a data warehouse

A data warehouse is a system that aggregates data from different sources into a single, central, consistent data store to support business analytics, data mining, Artificial Intelligence (AI) and machine learning. A data warehouse enables an organization to run powerful analytics on huge volumes (petabytes) of historical data in ways that a standard database cannot.

Data warehouses have been a part of Business Intelligence (BI) solutions for over three decades, but they have evolved significantly in recent years. Traditionally, a data warehouse was hosted on-premises (often on a mainframe computer) and its functionality was focused on extracting data from other sources, cleansing and preparing the data, loading and maintaining the data in a relational store. More recently, a data warehouse might be hosted on a dedicated appliance or in the cloud, and most data warehouses have added analytics capabilities, data visualization and presentation tools.

### 2. Benefits of a data warehouse

A data warehouse provides a foundation for the following.
- More consistent, higher-quality data: A data warehouse brings together data from multiple different sources, then cleanses it, eliminates duplicates and standardizes it to create a single source of the truth.
- Faster, unlimited insight: Disparate data sources limit the data that can be used to support any given decision. A data warehouse can easily provide decision support for companies.
- Smarter decision-making supported by cutting-edge tools: A data warehouse supports large-scale BI functions such as data mining (finding unseen patterns and relationships in data), artificial intelligence and machine learning—tools data professionals and business leaders can use to get hard evidence for making smarter decisions in virtually every area of the organization, from business processes to financial management and inventory management.
- Gaining and growing competitive advantage: Combine all of the above to help an organization to find more opportunities in data more quickly than that from disparate data stores.

### 3. Data warehouse architecture

Generally speaking, data warehouses have a three-tier architecture.
- The extraction tier collects, cleanses and normalizes/transforms the data from multiple sources by using a process known as ETL (extract, transform, load) or a process known as

ELT (extract, load, transform).

- The data store tier is typically a relational data store, but with schemas that support analytical processing.
- The analytics tier (or client layer) can include everything from standard querying tools to analytics, data mining, AI or machine learning capabilities and presentation visualization tools.

ETL and ELT are methods for extracting data from its original source and integrating it into the data warehouse. The difference between the two lies in where the data is transformed.

ETL extracts data from various data source systems, transforms it using an intermediate transformation engine, and then loads it into the data warehouse system. Because ETL transforms data before writing it to the warehouse, it's a better choice for loading smaller data volumes and for on-premises data warehouse solutions.

ELT extracts data from one or multiple remote sources and then loads it into the target data warehouse without any other formatting. The transformation of data in an ELT process happens within the target database. For this reason, ELT moves data to the warehouse faster, making it a better choice for larger data volumes or cloud-based data warehouse solutions. Also, because it doesn't transform data in transit, ELT is the only method suitable for loading a data lake.

**4. Data warehouse vs. database, data lake and data mart**

Confusion often arises between the terms data warehouse, database, data lake and data mart. Although the terms are similar, important differences exist.

(1) Data warehouse vs. data lake

A data warehouse gathers data from multiple sources into a central repository, structured using predefined schemas designed for data analytics. A data lake is basically a data warehouse without the predefined schemas. As a result, it enables more types of analytics than a data warehouse. Data lakes are commonly built on big data platforms such as Apache Hadoop.

(2) Data warehouse vs. data mart

A data mart is a subset of a data warehouse that contains data specific to a particular business line or department. Because they contain a smaller subset of data, data marts enable a department or business line to discover more-focused insights more quickly than possible when working with the broader data warehouse data set.

(3) Data warehouse vs. database

A database is built primarily for fast queries and transactional processing, not analytics. A database typically serves as the focused data store for a specific application, whereas a data warehouse stores data from any number (or even all) of the applications in your organization.

It is also important, while a database captures and stores data from a single (usually current) point in time, a data warehouse encompasses current and historical data required for predictive analytics, machine learning and other advanced analysis.

(4) Cloud data warehouse

A cloud data warehouse is a data warehouse specifically built to run in the cloud, and it is offered to customers as a managed service. Cloud-based data warehouses have grown more popular over the last five to seven years as more companies use cloud services and seek to reduce their on-premises data center footprint.

With a cloud data warehouse, the physical data warehouse infrastructure is managed by the cloud company, meaning that the customer doesn't have to make an upfront investment in hardware or software nor does he have to manage or maintain the data warehouse solution.

**5. Data warehouse software (on-premises/license)**

A business can purchase a data warehouse license and then deploy a data warehouse on their own on-premises infrastructure. Although this is typically more expensive than a cloud data warehouse service, it might be a better choice for government entities, financial institutions or other organizations that want more control over their data, need to comply with strict security, data privacy standards or regulations.

**6. Data warehouse appliance**

A data warehouse appliance is a pre-integrated bundle of hardware and software (CPUs, storage, operating system and data warehouse software) that a business can connect to its network and start using as-is. A data warehouse appliance sits somewhere between cloud and local facilities in terms of upfront cost, speed of deployment, ease of scalability and management control.

## ✎ New Words

| | | |
|---|---|---|
| aggregate | ['ægrɪgət] | n.合计；聚集体 |
| | | adj.总计的；聚合的 |
| | ['ægrɪgeɪt] | vt.总计，合计；使聚集，使积聚 |
| mainframe | ['meɪnfreɪm] | n.主机；大型机 |
| functionality | [ˌfʌŋkʃə'næləti] | n.功能性 |
| prepare | [prɪ'peə] | vt.准备；预备 |
| load | [ləʊd] | v.加载；装载 |
| | | n.负荷；装载；工作量 |
| eliminate | [ɪ'lɪmɪneɪt] | vt.排除，消除 |
| standardize | ['stændədaɪz] | vt.使标准化 |
| disparate | ['dɪspərət] | adj.完全不同的，根本不同的 |
| decision | [dɪ'sɪʒn] | n.决定 |
| cutting-edge | [ˌkʌtɪŋ'edʒ] | adj.前沿的 |
| pattern | ['pætn] | n.模式 |
| evidence | ['evɪdəns] | n.证据；迹象 |
| | | vt.显示；表明；证实 |
| tier | [tɪə] | n.级，阶，层；阶层，等级 |
| normalize | ['nɔ:məlaɪz] | vt.使规范化；使正常化；使标准化 |

| | | |
|---|---|---|
| transformation | [ˌtrænsfəˈmeɪʃn] | n.变化；转换；变换 |
| gather | [ˈɡæðə] | vt.收集；采集 |
| predefine | [ˈpriːdɪˈfaɪn] | vt.预定义；预先确定 |
| capture | [ˈkæptʃə] | vt.&n.捕捉 |
| encompass | [ɪnˈkʌmpəs] | vt.包含或包括某事物；完成 |
| seek | [siːk] | vt.寻找，探寻 |
| upfront | [ˌʌpˈfrʌnt] | adj.在前面的；预先的 |
| license | [ˈlaɪsns] | n.执照，许可证；特许 |
| | | vt.发许可证 |
| bundle | [ˈbʌndl] | n.捆；一批 |

## ✍ Phrases

| | |
|---|---|
| be hosted on | 被托管在……之上 |
| financial management | 财务管理 |
| transformation engine | 转换引擎 |
| target database | 目标数据库 |
| data lake | 数据湖 |
| data mart | 数据集市 |
| historical data | 历史数据 |
| cloud data warehouse | 云数据仓库 |
| operating system | 操作系统 |

## ✍ Abbreviations

| | |
|---|---|
| AI (Artificial Intelligence) | 人工智能 |
| ETL (Extract, Transform, Load) | 提取、转换、加载 |
| ELT (Extract, Load, Transform) | 提取、加载、转换 |
| CPU (Central Processing Unit) | 中央处理器 |

## Reference Translation

# 数 据 仓 库

### 1. 什么是数据仓库

　　数据仓库是一种系统，它将不同来源的数据聚合到单一的、集中的、一致的数据存储中，以支持业务分析、数据挖掘、人工智能（AI）和机器学习。数据仓库使组织能够以标准数据库无法实现的方式对大量（PB 级别）历史数据进行强大的分析。

数据仓库成为商业智能（BI）解决方案的一部分已经有三十多年的历史了，但是近年来它有了巨大的发展。传统上，数据仓库部署在本地——通常在大型主机上，其功能主要集中在从其他来源提取数据、清理和准备数据以及在关系存储中加载和维护数据。最近，数据仓库可能托管在专用设备上或云中，并且大多数的数据仓库都已经添加了分析功能以及数据可视化和演示工具。

**2. 数据仓库的好处**

数据仓库为以下各项提供了基础。

- 更一致、更高质量的数据：数据仓库将来自多个不同来源的数据汇集在一起，然后对其进行清理、消除重复并对其进行标准化以创建单一可信数据源。
- 更快、无限的洞察力：不同的数据源限制了可用于支持任何给定决策的数据。数据仓库可以轻松地为公司提供决策支持。
- 尖端工具支持更明智的决策：数据仓库支持大规模的商业智能功能，如数据挖掘（发现数据中看不见的模式和关系）、人工智能和机器学习——数据专业人员和业务领导者可以使用工具来获得可靠的证据以做出更明智的决策，其范围覆盖从企业流程到财务管理和库存管理的每个领域。
- 获得并增强竞争优势：结合以上所有优势，可以帮助组织更快地从数据中找到更多的机会，优于来自分散存储的数据。

**3. 数据仓库结构**

一般来说，数据仓库具有三层结构。

- 提取层使用称为 ETL（提取、转换和加载）的过程或称为 ELT（提取、加载和转换）的过程来收集、清理和规范化/转换来自多个源的数据。
- 数据存储层通常是一个关系数据存储库，但是具有支持分析处理的模式。
- 分析层（或客户层）可以包括从标准查询工具到分析、数据挖掘、人工智能或机器学习功能以及演示可视化工具的所有内容。

ETL 和 ELT 是从原始来源提取数据并将其集成到数据仓库中的方法。两者的区别在于数据转换的位置。

ETL 从各种数据源系统中提取数据，使用中间转换引擎对其进行转换，然后将其加载到数据仓库系统中。因为 ETL 在将数据写入仓库之前先进行数据转换，所以它是加载较小数据量和本地数据仓库解决方案的更好选择。

ELT 从一个或多个远程源中提取数据，然后将其加载到目标数据仓库中，而无须进行任何格式转换。ELT 流程中的数据转换发生在目标数据库中。因此，ELT 可以更快地将数据移至仓库，从而使其成为较大数据量或基于云的数据仓库解决方案的更好选择。此外，由于 ELT 不会转换传输中的数据，因此它是唯一适合加载数据湖的方法。

**4. 数据仓库与数据库、数据湖和数据集市**

在数据仓库、数据库、数据湖和数据集市之间经常会产生混淆。虽然这些术语相似，但存在重要区别。

（1）数据仓库与数据湖

数据仓库将来自多个源的数据收集到一个中央存储库中，该存储库使用为数据分析设计的预定义模式进行了结构化。数据湖基本上是没有预定义模式的数据仓库。因此，与数据仓库相比，它支持更多类型的分析。数据湖通常建立在像 Apache Hadoop 这样的大数据平台之上。

（2）数据仓库与数据集市

数据集市是数据仓库的子集，其中包含针对特定业务线或部门的数据。因为它们包含较小的数据子集，所以数据集市使部门或业务线能够比使用更广泛的数据仓库数据集时，更快地发现更侧重的见解。

（3）数据仓库与数据库

建立数据库主要是为了快速查询和事务处理，而不是分析。数据库通常用作特定应用程序的重点数据存储，而数据仓库则存储组织中任意数量的（甚至所有的）应用程序的数据。

同样重要的是，数据库从单个（通常是当前）时间点捕获并存储数据，而数据仓库包含预测分析、机器学习和其他高级分析所需的当前与历史数据。

（4）云数据仓库

云数据仓库是专门为在云中运行而构建的数据仓库，它作为托管服务提供给客户。在过去的 5～7 年中，随着越来越多的公司使用云服务并寻求减少其内部数据中心的占用空间，基于云的数据仓库变得越来越流行。

使用云数据仓库，物理数据仓库基础架构由云公司管理，这意味着客户不必在硬件或软件上进行前期投资，也不必管理或维护数据仓库解决方案。

**5. 数据仓库软件（本地/许可证）**

企业可以购买数据仓库许可证，然后在自己的本地基础结构上部署数据仓库。尽管这通常比云数据仓库服务昂贵，但对于希望更好地掌控数据或需要遵守严格的安全性或数据隐私标准或法规的政府实体、金融机构或其他组织，这可能是一个更好的选择。

**6. 数据仓库设备**

数据仓库设备是硬件和软件（CPU、存储、操作系统和数据仓库软件）的预集成捆绑包，企业可以将其连接到其网络并按原样使用。就前期成本、部署速度、易扩展性和管理控制而言，数据仓库设备介于云和本地设施之间。

# Exercises

**[Ex. 1]** 根据 **Text A** 回答以下问题。

1. What is a database?

2. What is SQL?

3. What does SQL consist of?

4. When did relational databases become dominant? What are items in a relational database organized as?

5. What does a distributed database consist of? Where may the database be stored?

6. What is a cloud database? What are the two types of cloud database models?

7. What is the newest and most groundbreaking type of database? What does it do?

8. What should general-purpose DBMS features and capabilities include?

9. Why is MySQL a popular choice for e-commerce businesses that need to manage multiple money transfers?

10. What are some common challenges that database administrators face?

[Ex. 2] 根据 Text B 回答以下问题。
1. What is a data warehouse?

2. Where was a data warehouse hosted traditionally? What is its functionality focused on?

3. What does a data warehouse provide a foundation for?

4. What are the three tiers data warehouses have generally speaking?

5. Why is ETL a better choice for loading smaller data volumes and for on-premises data warehouse solutions?

6. Where does the transformation of data in an ELT process happen?

7. What is a data lake? Where are data lakes commonly built on?

8. Why do data marts enable a department or business line to discover more-focused insights more quickly than possible when working with the broader data warehouse data set?

9. What is a cloud data warehouse?

10. What is a data warehouse appliance?

[Ex. 3] 词汇英译中
1. back end          1. _____
2. data lake          2. _____
3. cloud database     3. _____
4. data mart          4. _____

5. data access control      5. _____

6. calculus      6. _____

7. data warehouse      7. _____

8. concurrency      8. _____

9. distributed database      9. _____

10. control      10. _____

## [Ex. 4] 词汇中译英

1. 图形数据库      1. _____

2. 分布，分配；散发，分发      2. _____

3. 机器学习      3. _____

4. 引擎，发动机      4. _____

5. 非关系数据库      5. _____

6. 插入；嵌入      6. _____

7. 面向对象数据库      7. _____

8. 指令      8. _____

9. 面向对象程序设计      9. _____

10. 界面；接口      10. _____

## [Ex. 5] 短文翻译

### Data Lake

A data lake is a central location that holds a large amount of data in its native, raw format, as well as a way to organize large volumes of highly diverse data. Compared to a hierarchical data warehouse which stores data in files or folders, a data lake uses a different approach, it uses a flat architecture to store the data.

#### 1. Data lakes support all data types

A data lake holds big data from many sources in a raw, granular format. It can store structured, semi-structured or unstructured data, which means data can be kept in a more flexible format so we can transform it when we're ready to use it.

#### 2. Benefits of a data lake

Each element in a data lake gets assigned a unique identifier and is tagged with a set of extended metadata tags. Whenever there is a business question risen, the data lake can be queried for relevant data, and that smaller set of data can then be analyzed to help answer the question. You can apply various types of analytics to your data such as SQL queries, big data analytics, full-text search, real-time analytics, even machine learning can be used to uncover insights. Data lakes are usually configured on a cluster of scalable commodity hardware. As a result, data can be dumped in the lake in case it will be needed at a future date without worrying about storage capacity. In addition, the clusters could exist on-premises or in the cloud. The term data lake is usually associated with Hadoop oriented object storage.

# Reading Material

## What Is a Data Mart

In a market dominated by big data and analytics, data marts[1] are one key to efficiently transforming information into insights. Data warehouses typically deal with large data sets, but data analysis requires easy to find and readily available data. Should a business person have to perform complex queries just to access the data he need for reporting? No. And that's why smart companies use data marts.

A data mart is a subject oriented[2] database that is often a partitioned segment of an enterprise data warehouse. The subset of data held in a data mart typically aligns with a particular business unit like sales, finance or marketing. Data marts accelerate[3] business processes by allowing access to relevant information in a data warehouse or operational data store within days, as opposed to months or longer. Because a data mart only contains the data applicable to a certain business area, it is a cost-effective way to gain actionable insights quickly.

### 1. Data mart and data warehouse

Data marts and data warehouses are both highly structured repositories where data is stored and managed until it is needed. However, they differ in the scope of data stored: data warehouses are built to serve as the central store of data for the entire business, whereas a data mart fulfills[4] the request of a specific division or business function. Because a data warehouse contains data for the entire company, it is best practice to have strictly[5] control over who can access it. Additionally, querying the data you need in a data warehouse is an incredibly[6] difficult task for the business. Thus, the primary purpose of a data mart is to isolate or partition a smaller set of data from a whole to provide easier data access for the end consumers.

A data mart can be created from an existing data warehouse (using the top-down approach[7]) or from other sources, such as internal operational systems or external data. Similar to a data warehouse, it is a relational database that stores transactional data (time value, numerical order, reference to one or more objects) in columns and rows making it easy to organize and access.

Moreover, separate business units may create their own data marts based on their own data requirements. If business needs dictate, multiple data marts can be merged together to create a single

---

1   data mart: 数据集市

2   subject oriented: 面向主题

3   accelerate [ək'seləreɪt] vt.（使）加快，（使）增速

4   fulfill [fʊl'fɪl] vt.执行

5   strictly ['strɪktli] adv.严格地；完全地

6   incredibly [ɪn'kredəbli] adv.难以置信地，非常地

7   top-down approach: 自上而下的方法

data warehouse. This is the bottom-up development approach. The comparison between data mart and data warehouse is shown in table 4-1.

Table 4-1   The comparison between data mart and data warehouse

| | Data Mart | Data Warehouse |
|---|---|---|
| Size | < 100 GB | 100 GB + |
| Subject | Single Subject | Multiple Subjects |
| Scope | Line-of-Business | Enterprise-wide |
| Data Sources | Few Sources | Many Source Systems |
| Data Integration | One Subject Area | All Business Data |
| Time to Build | Minutes, Weeks, Months | Many Months to Years |

## 2. Types of data marts

There are three types of data marts: dependent, independent and hybrid. They are categorized based on their relation to the data warehouse and the data sources that are used to create the system.

(1) Dependent data marts

A dependent data mart is created from an existing enterprise data warehouse. It is the top-down approach that begins with storing all business data in one central location, then extracts a clearly defined portion of the data when needed for analysis.

To form a data warehouse, a specific set of data is aggregated (formed into a cluster) from the warehouse, restructured[8], then loaded to the data mart where it can be queried. It can be a logical view or physical subset of the data warehouse.

- Logical view[9]. A virtual table/view that is logically (but not physically) separated from the data warehouse.
- Physical subset. Data extract that is a physically separate database from the data warehouse.

(2) Independent data marts

An independent data mart is a stand-alone system, created without the use of a data warehouse, that focuses on one subject area or business function. Data is extracted from internal or external data sources (or both), processed, then loaded to the data mart repository where it is stored until needed for business analytics.

Independent data marts are not difficult to design and develop. They are beneficial to achieve short-term goals but may become cumbersome[10] to manage, each with its own ETL tools and logic, as business needs expand and become more complex.

(3) Hybrid data marts

A hybrid data mart combines data from an existing data warehouse and other operational source systems. It unites[11] the speed and end-user focus of a top-down approach with the benefits of

---

8   restructure [ˌriːˈstrʌktʃə] v.重建，重组；调整

9   logical view: 逻辑视图

10   cumbersome [ˈkʌmbəsəm] adj.笨重的；累赘的；冗长的；麻烦的

11   unite [juˈnaɪt] v.（使）联合

the enterprise-level integration of the bottom-up method.

## 3. Structure of a data mart

Similar to a data warehouse, a data mart may be organized using a star, snowflake or other schema as a blueprint. IT teams typically use a star schema consisting of one or more fact tables[12] (set of metrics relating to a specific business process or event) referencing dimension tables[13] (primary key joined to a fact table) in a relational database.

The benefit of a star schema is that fewer joins are needed when writing queries, as there is no dependency between dimensions. This simplifies the ETL request process, making it easier for analysts to access and navigate.

In a snowflake schema, dimensions are not clearly defined. They are normalized to help reduce data redundancy and protect data integrity. It takes less space to store dimension tables, but it is a more complicated structure (multiple tables to populate[14] and synchronize) that can be difficult to maintain.

## 4. Advantages of a data mart

Managing big data and gaining valuable business insights is a challenge that all companies face, and one that most are answering with strategic data marts.

- Efficient access. A data mart is a time-saving solution for accessing a specific set of data for business intelligence[15].
- Inexpensive data warehouse alternative[16]. Data marts can be an inexpensive alternative to developing an enterprise data warehouse, where required data sets are smaller. An independent data mart can be up and running in a week or less.
- Improve data warehouse performance. Dependent and hybrid data marts can improve the performance of a data warehouse by taking on the burden of processing to meet the needs of the analyst. When dependent data marts are placed in a separate processing facility, they significantly reduce analytics processing costs as well.

Other advantages of a data mart are as follows.

- Data maintenance. Different departments can own and control their data.
- Simple setup. The simple design requires less technical skill to set up.
- Analytics. Key Performance Indicators (KPI)[17] can be easily tracked.
- Easy entry. Data marts can be the building blocks of a future enterprise data warehouse project.

## 5. The future of data marts

Even with the improved flexibility and efficiency that data marts offer, big data is still

---

12   fact table: 事实表

13   dimension table: 维度表

14   populate ['pɒpjuleɪt] *vt.*填充

15   business intelligence: 商业智能

16   alternative [ɔːl'tɜːnətɪv] *n.*可供选择的事物 *adj.*替代的，备选的

17   Key performance indicators (KPIs): 关键绩效指标

becoming too big for many on-premises solutions. As data warehouses and data lakes move to the cloud, so too do data marts.

With a shared cloud-based platform to create and house data, access and analytics become much more efficient. Transient[18] data clusters can be created for short-term analysis, or long-lived clusters can come together for more sustained[19] work. Modern technologies are also separating data storage from compute, allowing for ultimate scalability for querying data.

Other advantages of cloud-based dependent and hybrid data marts are as follows.

- Flexible architecture with cloud-native applications[20].
- Single depository containing all data marts.
- Resources consumed on-demand.
- Immediate real-time access to information.
- Increased efficiency.
- Consolidation of resources that lowers costs.
- Real-time and interactive analytics.

---

18  transient ['trænzɪənt] *adj.*短暂的；转瞬即逝的；临时的
19  sustained [səs'teɪnd] *adj.*持久的，持续的
20  cloud-native application: 云原生应用程序

# Unit 5

## Text A

## ETL

扫码听课文

### 1. What is ETL

ETL is a process that extracts the data from different source systems, then transforms the data (like applying calculations, concatenations, etc.) and finally loads the data into the data warehouse system. The full form of ETL is Extract, Transform and Load.

You will think creating a data warehouse is simply extracting data from multiple sources and loading into database of a data warehouse. This is far from the truth and requires a complex ETL process. The ETL process requires active inputs from various stakeholders including developers, analysts, testers, top executives and is technically challenging.

In order to maintain its value as a tool for decision-makers, data warehouse system needs to change with business changes. ETL is a recurring activity (daily, weekly, monthly) of a data warehouse system and needs to be agile, automated and well documented.

### 2. Why do you need ETL

There are many reasons for adopting ETL in the organization.

- It helps companies to analyze their business data for making critical business decisions.
- It can answer complex business questions that transactional databases cannot.
- It provides a method of moving the data from various sources into a data warehouse.
- Well-designed and documented ETL system is almost essential to the success of a data warehouse project.
- It allows verification of data transformation, aggregation and calculations rules.
- ETL process allows sample data comparison between the source and the target system.
- ETL process can perform complex transformations and requires the extra area to store the data.
- It helps to migrate data into a data warehouse and convert it to the various formats and types to adhere to one consistent system.

### 3. ETL process in data warehouses

Step 1. Extraction

In this step, data is extracted from the source system and stored into the staging area. Transformations, if any, are done in staging area so that performance of source system is not degraded. Also, if corrupted data is copied directly from the source into data warehouse, rollback will be a challenge. Staging area gives an opportunity to validate extracted data before it moves into the data warehouse.

Data warehouse needs to integrate systems that have different DBMS, hardware, operating systems and communication protocols. Sources could include legacy applications like mainframes, customized applications, point of contact devices like ATM, call switches, text files, spreadsheets, ERP and data from vendors or partners among others.

Hence one needs a logical data map before data is extracted and loaded physically. This data map describes the relationship between sources and target data.

There are three data extraction methods.

- Full extraction.
- Partial extraction: without update notification.
- Partial extraction: with update notification.

Irrespective of the method used, extraction should not affect performance and response time of the source systems. These source systems are live production databases. Any slow down or locking could effect company's bottom line.

Some validations are done during extraction, as follows.

- Reconcile records with the source data.
- Make sure that no spam/unwanted data is loaded.
- Check data type.
- Remove all types of duplicate data.
- Check whether all the keys are in place or not.

Step 2. Transformation

Data extracted from source server is raw and not usable in its original form. Therefore, it needs to be cleansed, mapped and transformed. In fact, this is the key step where ETL process adds value and changes data such that insightful BI reports can be generated.

In this step, you apply a set of functions on extracted data. Data that does not require any transformation is called direct move.

In transformation step, you can perform customized operations on data. For instance, if the user wants sum-of-sales revenue which is not in the database. Or if the first name and the last name in a table is in different columns. It is possible to concatenate them before loading.

Following are data integrity problems.

- There are different spelling of the same person like Jon, John, etc.
- There are multiple ways to denote company name like Google, Google Inc.
- There are different names like Cleaveland, Cleveland.
- There may be a case that different account numbers are generated by various applications for the same customer.
- Some data required in files remains blank.

Complete the following validations at this stage.

- Filtering: select only certain columns to load.

- Using rules and lookup tables for data standardization.
- Character set conversion and encoding handling.
- Conversion of units of measurements like date time conversion, currency conversions, numerical conversions, etc.
- Data threshold validation check. For example, age cannot be more than two digits.
- Data flow validation from the staging area to the intermediate tables.
- Required fields should not be left blank.
- Cleaning, for example, mapping NULL to 0 or Gender Male to "M" and Female to "F" etc.
- Splitting a column into multiples and merging multiple columns into a single column.
- Transposing rows and columns.
- Using lookups to merge data.
- Using any complex data validation. For example, if the first two columns in a row are empty then it automatically reject the row from processing.

Step 3. Loading

Loading data into the target data warehouse is the last step of the ETL process. In a typical data warehouse, huge volume of data needs to be loaded in a relatively short period. Hence, load process should be optimized for performance.

In case of load failure, recover mechanisms should be configured to restart from the point of failure without data integrity loss. Data warehouse admins need to monitor, resume or cancel loads as per prevailing server performance.

There are three types of loading, as follows.

- Initial load: populating all the data warehouse tables.
- Incremental load: applying ongoing changes when needed periodically.
- Full refresh: erasing the contents of one or more tables and reloading with fresh data.

Load verification as follows.

- Ensure that the key field data is neither missing nor null.
- Test modeling views based on the target tables.
- Check combined values and calculated measures.
- Check data in dimension table as well as history table.
- Check the BI reports on the loaded fact and dimension table.

## 4. ETL tools

There are many data warehousing tools are available in the market. Here are some most prominent ones.

(1) MarkLogic

MarkLogic is a data warehousing solution which makes data integration easier and faster using an array of enterprise features. It can query different types of data like documents, relationships and metadata.

(2) Oracle

Oracle is the industry leading database. It offers a wide range of choice of data warehouse solutions for both on-premises and in the cloud. It helps to optimize customer experiences by increasing operational efficiency.

(3) Amazon RedShift

Amazon Redshift is data warehouse tool. It is a simple and cost-effective tool to analyze all types of data using standard SQL and existing BI tools. It also allows running complex queries against petabytes of structured data.

**5. Best practices of ETL process**

(1) Never try to cleanse all the data

Every organization would like to have all the data clean, but most of them are not ready to pay to wait or not ready to wait. To clean it all would simply take too long, so it is better not to try to cleanse all the data.

(2) Plan to clean something

Always plan to clean something because the biggest reason for building the data warehouse is to offer cleaner and more reliable data.

(3) Determine the cost of cleansing the data

Before cleansing all the dirty data, it is important for you to determine the cleansing cost for every dirty data element.

(4) Store summarized data into disk tapes

To reduce storage costs, store summarized data into disk tapes. Also, the trade-off between the volume of data to be stored and its detailed usage is required. Trade off at the level of granularity of data to decrease the storage costs.

**6. Summary**

- ETLstands for Extract, Transform and Load.
- ETL provides a method of moving the data from various sources into a data warehouse.
- In the extraction step, data is extracted from the source system into the staging area.
- In the transformation step, the data extracted from source is cleansed and transformed .
- Loading data into the target data warehouse is the last step of the ETL process.

## New Words

| | | |
|---|---|---|
| concatenation | [kən͵kætə'neɪʃn] | n.互相关联的事物 |
| complex | ['kɒmpleks] | adj.复杂的；复合的 |
| stakeholder | ['steɪkhəʊldə] | n.股东；利益相关者 |
| analyst | ['ænəlɪst] | n.分析家，分析师 |
| tester | ['testə] | n.测试员 |
| recur | [rɪ'kɜː] | vi.再发生；复发；重现 |
| agile | ['ædʒaɪl] | adj.敏捷的，灵活的 |
| adopt | [ə'dɒpt] | v.采用（某方法）；采取（某态度） |
| critical | ['krɪtɪkl] | adj.关键的；严重的；极重要的 |

| | | |
|---|---|---|
| project | ['prɒdʒekt] | n.项目；方案；工程，计划 |
| verification | [ˌverɪfɪ'keɪʃn] | n.验证；证明；证实 |
| aggregation | [ˌægrɪ'geɪʃn] | n.聚集；集成；集结 |
| sample | ['sɑ:mpl] | n.样本；样品 |
| | | vt.抽样调查；取样 |
| comparison | [kəm'pærɪsn] | n.比较，对照 |
| extra | ['ekstrə] | adj.额外的，补充的，附加的 |
| codify | ['kəʊdɪfaɪ] | vt.编纂，整理；编成法典 |
| reuse | [ˌri:'ju:z] | vt.再用，重新使用 |
| degrade | [dɪ'greɪd] | vt.降低，降级 |
| corrupt | [kə'rʌpt] | adj.损坏的，（文献等）错误百出的 |
| copy | ['kɒpi] | v.复制 |
| rollback | ['rəʊlbæk] | n.回滚 |
| map | [mæp] | v.映射 |
| notification | [ˌnəʊtɪfɪ'keɪʃn] | n.通知；布告；公布 |
| irrespective | [ɪrɪ'spektɪv] | adj.不考虑的，不顾的；无关的 |
| reconcile | ['rekənsaɪl] | vt.使一致 |
| spam | [spæm] | n.垃圾邮件 |
| duplicate | ['dju:plɪkeɪt] | v.复制 |
| | | adj.复制的；副本的 |
| | | n.复制品；副本 |
| usable | ['ju:zəbl] | adj.可用的；合用的；便于使用的 |
| insightful | ['ɪnsaɪtfʊl] | adj.富有洞察力的，有深刻见解的 |
| filtering | ['fɪltərɪŋ] | v.过滤 |
| lookup | ['lʊkʌp] | v.查找；查表 |
| conversion | [kən'vɜ:ʃn] | n.变换，转变 |
| threshold | ['θreʃhəʊld] | n.门槛；阈值；临界值 |
| intermediate | [ˌɪntə'mi:diət] | adj.中间的，中级的 |
| | | n.中间物 |
| merge | [mɜ:dʒ] | v.（使）混合；相融；融入 |
| empty | ['empti] | adj.空的 |
| reject | [rɪ'dʒekt] | vt.拒绝；抛弃，扔掉 |
| failure | ['feɪljə] | n.失败 |
| recover | [rɪ'kʌvə] | vt.恢复；重新获得；找回 |

| | | |
|---|---|---|
| restart | [ˌriːˈstɑːt] | v.重新开始 |
| prevailing | [prɪˈveɪlɪŋ] | adj.占优势的；主要的；普遍的 |
| refresh | [rɪˈfreʃ] | vt.刷新；使恢复 |
| erase | [ɪˈreɪz] | vt.擦掉；抹去；清除 |
| combine | [kəmˈbaɪn] | v.使结合；组合 |
| reliable | [rɪˈlaɪəbl] | adj.可靠的；可信赖的 |
| summarize | [ˈsʌməraɪz] | vt.总结，概述 |
| granularity | [ˌgrænjəˈlærəti] | n.间隔尺寸，粒度 |

## ✍ Phrases

| | |
|---|---|
| in order to | 为了…… |
| business data | 业务数据，商业数据 |
| business decision | 业务决策，商业决策 |
| sample data | 样本数据 |
| adhere to | 遵循；依附；坚持 |
| technical skill | 技术技能；专门技能 |
| staging area | 暂存区域，临时区域 |
| integrate system | 整合系统，集成系统 |
| communication protocol | 通信协议 |
| call switch | 呼叫交换 |
| text file | 文本文件 |
| data map | 数据映射 |
| response time | 响应时间 |
| bottom line | 最终盈利 |
| direct move | 直接移动 |
| data integrity | 数据完整性 |
| account number | 账号 |
| lead to | 导致 |
| in a short period | 短期内 |
| be configured to | 被配置为 |
| customer experience | 客户体验 |
| trade off | 权衡，交易 |

## ✍ Abbreviations

| | |
|---|---|
| BI (Business Intelligence) | 商务智能 |
| POS (Point of Sale) | 销售终端 |

## Reference Translation

## 提取、转换和加载

### 1. 什么是 ETL

ETL 是一个从不同源系统提取数据，然后转换数据（如应用计算、连接等），最后将数据加载到数据仓库系统的过程。ETL 的完整形式是提取、转换和加载。

有人认为创建数据仓库就只是从多个来源提取数据并将其加载到数据仓库的数据库中。这远非事实，它需要复杂的 ETL 过程。ETL 流程需要包括开发人员、分析师、测试人员、高层管理人员在内的各种利益相关者的积极投入，这在技术上具有挑战性。

为了保持其作为决策者工具的价值，数据仓库系统需要随业务变化而变化。ETL 是数据仓库系统的经常性活动（每天、每周、每月），并且需要敏捷、自动并且文档完备。

### 2. 为什么需要 ETL

在组织中采用 ETL 的原因有很多，具体如下。

- 它可以帮助公司分析其业务数据以做出关键的业务决策。
- 它可以回答事务数据库无法回答的复杂业务问题。
- 它提供了一种将数据从各种来源移到数据仓库中的方法。
- 设计良好且文档完备的 ETL 系统对于数据仓库项目的成功至关重要。
- 它允许验证数据转换、聚合和计算规则。
- ETL 过程允许在源系统和目标系统之间进行样本数据比较。
- ETL 过程可以执行复杂的转换，并且需要额外的区域来存储数据。
- 它有助于将数据迁移到数据仓库中并将其转换为各种格式和类型，以保持系统的一致性。

### 3. 数据仓库中的 ETL 过程

步骤 1. 提取

在此步骤中，从源系统提取数据并存储到暂存区域中。转换（如果有的话）在暂存区域中进行，因此源系统的性能不会降低。此外，如果将损坏的数据直接从源复制到数据仓库中，回滚将是一个挑战。暂存区提供了在提取的数据移入数据仓库之前对其进行验证的机会。

数据仓库需要集成具有不同 DBMS、硬件、操作系统和通信协议的系统。数据来源可能包括传统应用程序（如大型机）、定制应用程序、接触点设备（如 ATM）、呼叫交换、文本文件、电子表格、ERP 和来自供应商或合作伙伴等的数据。

因此，在物理上提取和加载数据之前，需要一个逻辑数据映射。该数据映射描述了源数据和目标数据之间的关系。

数据提取方法有以下三种。

- 全提取。
- 部分提取：没有更新通知。
- 部分提取：有更新通知。

无论使用哪种方法，提取都不应影响源系统的性能和响应时间。这些源系统实时产生数据库。任何放缓或锁定都可能影响公司的最终盈利。

在提取过程中会进行一些验证。

- 使记录与源数据一致。
- 确保没有垃圾邮件/不需要的数据被加载。
- 检查数据类型。
- 删除所有类型的重复数据。
- 检查所有关键数据到位与否。

步骤 2. 转换

从源服务器提取的数据是原始数据，不能以其原始形式使用。因此，需要对其进行清理、映射和转换。实际上，这是 ETL 流程的关键步骤，它增加了数据的价值且改变了数据，以生成具有洞察力的商务智能报告。

在此步骤中，你可以对提取的数据进行一些处理。不需要任何转换的数据称为直接移动。

在转换步骤中，你可以对数据执行自定义操作。例如，如果用户想要销售总额的收入，而该收入不在数据库中。或者，如果表中的名字和姓氏在不同的列中。你可以在加载之前将它们关联起来。

以下是数据完整性问题。

- 同一个人，名字 Jon、John 拼写不同。
- 有多种表示公司名称的方法，如 Google、Google Inc.。
- 有不同的名称，如 Cleaveland、Cleveland。
- 可能会有不同的应用程序为同一位客户生成不同账号的情况。
- 所需文件中的某些数据是空的。

在此阶段完成以下验证。

- 过滤：仅选择某些列来加载。
- 使用规则和查找表进行数据标准化。
- 字符集转换和编码处理。
- 度量单位的转换，如日期时间转换、货币转换、数字转换等。
- 数据阈值验证检查。例如，年龄不超过两位数。
- 从暂存区域到中间表的数据流验证。
- 必填字段不应留为空白。
- 清理，例如，将 NULL 映射为 0 或将 "Gender Male" 映射为 "M"，将 "Female" 映射为 "F" 等。
- 将一列拆分为多列以及将多个列合并为一列。
- 转置行和列。
- 使用查找合并数据。
- 使用任何复杂的数据验证。例如，如果一行中的前两列为空，那么自动拒绝对该行进行处理。

步骤 3. 加载

将数据加载到目标数据仓库是 ETL 过程的最后一步。在典型的数据仓库中，需要在相对较短的时间内加载大量数据。因此，应优化加载过程。

如果发生加载故障，应将恢复机制配置为从故障点重新启动，从而不会丢失数据完整性。数据仓库管理员需要根据当前服务器的性能来监管、恢复或取消加载。

有三种加载类型，具体如下。

- 初始加载：填充所有数据仓库表。
- 增量加载：当需要时，定期进行持续更改。
- 完全刷新：擦除一个或多个表的内容并重新加载新数据。

加载验证如下。

- 确保关键字段数据不丢失也不为空。
- 根据目标表测试建模视图。
- 检查组合值和计算结果。
- 检查维度表和历史记录表中的数据。
- 检查 BI 报告中已加载的事实和维度表。

## 4. ETL 工具

市场上有许多可用的数据仓库工具。此处罗列了其中一些最著名的。

（1）MarkLogic

MarkLogic 是一种数据仓库解决方案，可使用一系列企业功能使数据集成变得更加轻松快捷。它可以查询不同类型的数据，例如文档、关系和元数据。

（2）Oracle

Oracle 是行业领先的数据库。它为本地和云提供了广泛的数据仓库解决方案。它通过提高运营效率来帮助优化客户体验。

（3）Amazon RedShift

Amazon Redshift 是一种数据仓库工具。它是使用标准 SQL 和现有商务智能工具分析所有类型数据的简单且经济高效的工具。它还允许对 PB 字节级结构化数据运行复杂的查询。

## 5. ETL 过程的最佳实践

（1）永远不要尝试清理所有数据

每个组织都希望所有数据都是干净的，但是大多数组织不愿意支付等待的费用或不想等待。清理全部数据将花费很长时间，因此最好不要尝试清理所有数据。

（2）计划清理内容

始终制订清理内容计划，因为构建数据仓库的最大目的是提供更干净、更可靠的数据。

（3）确定清理数据的成本

在清理所有脏数据之前，确定每个脏数据元素的清理成本非常重要。

（4）将汇总数据存储到磁盘磁带中

为了降低存储成本，请将摘要数据存储到磁盘磁带中。而且，需要在要存储的数据量及其详细用法之间进行权衡。在数据的粒度级别上进行权衡以降低存储成本。

## 6. 总结

- ETL 代表提取、转换和加载。
- ETL 提供了一种将数据从各种来源转移到数据仓库中的方法。
- 在提取步骤中，将数据从源系统提取到暂存区中。
- 在转换步骤中，将从源中提取的数据进行清理和转换。

● 将数据加载到目标数据仓库是 ETL 过程的最后一步。

扫码听课文

# Text B

# Big Data Analytics Tools and Their Key Features

With the rise in the volume of big data and tremendous growth in cloud computing, the cutting edge big data analytics tools have become the key to achieve a meaningful analysis of data. In this article, we shall discuss the top big data analytics tools and their key features.

**1. Apache Storm**

Apache Storm is an open-source and free big data computation system. It is also an Apache product with a real-time framework for data stream processing which supports any programming language. It offers a distributed real-time, fault-tolerant processing system with real-time computation capabilities. Storm scheduler manages workload with multiple nodes with reference to topology configuration and it works well with the Hadoop Distributed File System (HDFS).

Features:

- It is benchmarked as processing one million 100 byte messages per second per node.
- Storm assures that unit of data will be processed at least once.
- It has great horizontal scalability.
- It has built-in fault-tolerance.
- It auto-restarts on crashes.
- It is Clojure written.
- It works with Directed Acyclic Graph (DAG) topology.
- Output files are in JSON format.
- It has multiple use cases—real-time analytics, log processing, ETL, continuous computation, distributed RPC, machine learning, etc.

**2. Talend**

Talend is a big data tool that simplifies and automates big data integration. Its graphical wizard generates native code. It also allows big data integration, master data management and checks data quality.

Features:

- It streamlines ETL and ELT for big data.
- It accomplishes the speed and scale of Spark.
- It accelerates your move to real-time.
- It handles multiple data sources.
- It provides numerous connectors, which in turn will allow you to customize the solution as per your need.
- Talend Big Data Platform simplifies using MapReduce and Spark by generating native code.
- It improves data quality with machine learning and natural language processing.

- Its agile DevOps can speed up big data projects.
- It streamline all the DevOps processes.

## 3. Apache CouchDB

Apache CouchDB is an open-source, cross-platform and document-oriented NoSQL database that aims at ease of use and holding a scalable architecture. It is written in concurrency-oriented language Erlang. CouchDB stores data in JSON documents that can be accessed through web or queried using JavaScript. It offers distributed scaling with fault-tolerant storage.

Features:

- CouchDB is a single-node database that works like any other database.
- It allows running a single logical database server on any number of servers.
- It makes use of the ubiquitous HTTP and JSON data format.
- Document insertion, updates, retrieval and deletion are quite easy.
- JSON format can be translatable across different languages.

## 4. Apache Spark

Apache Spark is a very popular and open-source big data analytics tool. Spark has many high-level operators for easily building parallel applications. It is used at a wide range of organizations to process large datasets.

Features:

- It helps to run an application in Hadoop cluster, up to 100 times faster in memory and ten times faster on disk.
- It offers fast processing.
- It supports sophisticated analytics.
- It can integrate with Hadoop and existing Hadoop Data.
- It provides built-in APIs in Java, Scala or Python.
- Spark provides the in-memory data processing capabilities, which is much faster than disk processing leveraged by MapReduce.
- Spark works with HDFS, OpenStack and Apache Cassandra, both in the cloud and on-premise, adding another layer of versatility to big data operations for your business.

## 5. Splice Machine

Splice Machine is a big data analytics tool. Their architecture is portable across public clouds such as AWS, Azure and Google.

Features:

- It can dynamically scale from a few to thousands of nodes to enable applications at every scale.
- The Splice Machine optimizer automatically evaluates every query to the distributed HBase regions.
- It can reduces management, deploys faster and reduces risk.
- It can deal with fast streaming data, develops, tests and deploys machine learning models.

## 6. Plotly

Plotly is an analytics tool that lets users create charts and dashboards to share online.

Features:

- It easily turns any data into eye-catching and informative graphics.
- It provides audited industries with fine-grained information on data provenance.
- It offers unlimited public file hosting through its free community plan.

## 7. Azure HDInsight

Azure HDInsight is a Spark and Hadoop service in the cloud. It provides big data cloud offerings in two categories, Standard and Premium. It provides an enterprise-scale cluster for the organization to run their big data workloads.

Features:

- It offers reliable analytics with an industry-leading SLA.
- It offers enterprise-grade security and monitoring.
- It protects data assets and extends on-premises security and governance controls to the cloud.
- It is a high-productivity platform for developers and scientists.
- It is integrated with leading productivity applications.
- It deploys Hadoop in the cloud without purchasing new hardware or paying other up-front costs.

## 8. Skytree

Skytree is a big data analytics tool that empowers data scientists to build more accurate models faster. It offers accurate predictive machine learning models that are easy to use.

Features:

- It has highly scalable algorithms.
- It is artificial intelligence for data scientists.
- It allows data scientists to visualize and understand the logic behind Machine Learning (ML) decisions.
- It is easy to adopt GUI or programmatically in Java via Skytree.
- It has model interpretability.
- It is designed to solve robust predictive problems with data preparation capabilities.
- It is programmatic and allows GUI access.

## 9. Lumify

Lumify is considered a visualization platform, big data fusion and analysis tool. It helps users to discover connections and explore relationships in their data via a suite of analytic options.

Features:

- It provides both 2D and 3D graph visualizations with a variety of automatic layouts.
- It provides link analysis between graph entities, integration with mapping systems, geospatial analysis, multimedia analysis, real-time collaboration through a set of projects or workspaces.
- It comes with specific ingest processing and interface elements for textual content, images and videos.
- Its space feature allows you to organize work into a set of projects or workspaces.

• It is built on proven, scalable big data technologies.

• It supports the cloud-based environment. It works well with Amazon's AWS.

## 10. Hadoop

Hadoop is the long-standing champion in the field of big data processing. It is well-known for its capabilities for huge-scale data processing. It has low hardware requirement due to open-source big data framework can run on-premise or in the cloud.

Features:

• It improves authentication when using HTTP proxy server.

• It supports POSIX-style file system extended attributes.

• It offers a robust ecosystem that is well suited to meet the analytical needs of a developer.

• It brings flexibility in data processing.

• It allows for faster data processing.

## New Words

| | | |
|---|---|---|
| fault-tolerant | ['fɔ:lt 'tɒlərənt] | adj.容错的 |
| scheduler | ['ʃedju:lə] | n.调度程序，日程安排程序 |
| workload | ['wɜ:kləud] | n.工作量，工作负担 |
| benchmark | ['bentʃmɑ:k] | n.基准，参照 |
| built-in | [ˌbɪlt' ɪn] | adj.嵌入的；内置的；固有的 |
| auto-restart | ['ɔ:təu ˌri:'stɑ:t] | n.自动重启 |
| topology | [tə'pɒlədʒi] | n.拓扑 |
| cross-platform | [krɒs 'plætfɔ:m] | adj.跨平台的 |
| document-oriented | ['dɒkjumənt 'ɔ:rɪəntɪd] | adj.面向文档的 |
| concurrency-oriented | [kən'kʌrənsɪ 'ɔ:rɪəntɪd] | adj.面向并发的 |
| ubiquitous | [ju:'bɪkwɪtəs] | adj.普遍存在的 |
| insertion | [ɪn'sɜ:ʃn] | n.插入 |
| translatable | [træns'leɪtəbl] | adj.可译的，可转换的 |
| parallel | ['pærəlel] | adj.并行的；平行的 |
| sophisticated | [sə'fɪstɪkeɪtɪd] | adj.复杂的；精致的；富有经验的 |
| versatility | [ˌvɜ:sə'tɪləti] | n.多用途 |
| region | ['ri:dʒən] | n.范围，领域 |
| dashboard | ['dæʃbɔ:d] | n.仪表板；仪表盘 |
| eye-catching | ['aɪ 'kætʃɪŋ] | adj.引人注目的；显著的 |
| fine-grained | ['faɪn 'greɪnd] | adj.精确的 |
| up-front | ['ʌp 'frʌnt] | adj.预付的 |
| empower | [ɪm'pauə] | vt.授权；准许；使能够 |
| visualize | ['vɪʒuəlaɪz] | vt.使形象化，使可视化 |
| programmatically | [ˌprəugrə'mætɪkli] | adv.编程地 |
| interpretability | [ɪn'tɜ:prɪtəbɪlɪti] | n.可解释性；解释能力；可解读性 |

| | | |
|---|---|---|
| fusion | ['fju:ʒn] | n.融合 |
| layout | ['leɪaʊt] | n.布局，安排，设计；层 |
| geospatial | [,dʒi:əʊ'speɪʃəl] | adj.地理空间的 |
| multimedia | [,mʌltɪ'mi:dɪə] | adj.多媒体的 |
| collaboration | [kə,læbə'reɪʃn] | n.合作，协作 |
| workspace | ['wɜ:kspeɪs] | n.工作区，工作空间 |
| ingest | [ɪn'dʒest] | vt.获取（某事物）；吸收 |
| proxy | ['prɒksi] | n.代理 |

## ✍ Phrases

| | |
|---|---|
| cutting edge | 尖端；最前沿；优势 |
| data stream | 数据流 |
| use case | 用例 |
| native code | 本机代码，本地代码 |
| natural language processing | 自然语言处理 |
| integrate with | 与……结合，与……集成 |
| visualization platform | 可视化平台 |
| a suite of | 一系列；一套 |
| proxy server | 代理服务器 |

## ✍ Abbreviations

| | |
|---|---|
| HDFS (Hadoop Distributed File System) | Hadoop 分布式文件系统 |
| RPC (Remote Procedure Call) | 远程过程调用 |
| HTTP (Hyper Text Transfer Protocol) | 超文本传输协议 |
| SLA (Service Level Agreement) | 服务等级协议，服务级别协议 |
| GUI (Graphical User Interface) | 图形用户界面，图形用户接口 |
| 2D (2-Dimensional) | 二维 |
| 3D (3-Dimensional) | 三维 |
| POSIX (Portable Operating System Interface of UNIX) | UNIX 可移植操作系统接口 |

## Reference Translation

## 大数据分析工具及其主要功能

随着大数据量的增加和云计算的巨大增长，尖端的大数据分析工具已成为实现有意义的数据分析的关键。在本文中，我们将讨论顶级的大数据分析工具及其主要功能。

**1. Apache Storm**

Apache Storm 是一个开源的免费大数据计算系统。它也是具有实时框架的 Apache 产品，用于支持任何编程语言的数据流处理。它提供一个具有实时计算功能的分布式的实时、容错处理系统。Storm 调度程序通过参考拓扑配置来管理具有多个节点的工作负载，并且可以与 Hadoop 分布式文件系统（HDFS）很好地配合使用。

功能：

- 以每个节点每秒处理一百万条 100 字节的消息为基准。
- Storm 确保数据单元至少被处理一次。
- 具有出色的水平可扩展性。
- 具有内置的容错功能。
- 崩溃时自动重新启动。
- 它是用 Clojure 语言编写的。
- 它与有向无环图（DAG）拓扑一起使用。
- 输出文件为 JSON 格式。
- 它具有多个用例：实时分析、日志处理、ETL、连续计算、分布式 RPC、机器学习。

**2. Talend**

Talend 是一个大数据工具，可简化和自动化执行大数据集成。它的图形向导可生成本机代码。它还允许进行大数据集成、主数据管理并检查数据质量。

功能：

- 简化大数据的 ETL 和 ELT。
- 实现 Spark 的速度和规模。
- 加快你的实时性。
- 处理多个数据源。
- 提供大量连接器，这又使得可以根据需要自定义解决方案。
- Talend 大数据平台通过生成本机代码简化了 MapReduce 和 Spark 的使用。
- 通过机器学习和自然语言处理提高数据质量。
- 其敏捷的 DevOps 可加速大数据项目。
- 简化所有 DevOps 流程。

**3. Apache CouchDB**

Apache CouchDB 是一个开放源代码、跨平台、面向文档的 NoSQL 数据库，旨在易于使用并拥有可扩展的体系结构。它是用面向并发的语言 Erlang 编写的。CouchDB 将数据存储在 JSON 文档中，可以通过网络进行访问或使用 JavaScript 查询。它提供带有容错存储的分布式扩展性能。

功能：

- CouchDB 是一个单节点数据库，其工作方式与任何其他数据库一样。
- 它允许在任意数量的服务器上运行单个逻辑数据库服务器。
- 它使用了无处不在的 HTTP 和 JSON 数据格式。
- 文档插入、更新、检索和删除非常容易。
- 可以在不同语言之间转移 JSON 格式。

## 4. Apache Spark

Apache Spark 是一个非常流行的开源大数据分析工具。Spark 拥有许多运行器，可轻松构建并行应用程序。它被广泛用于组织中以处理大型数据集。

功能：

- 它有助于在 Hadoop 集群中运行应用程序，把内存速度提高多达 100 倍，磁盘速度提高 10 倍。
- 提供快速处理。
- 它支持复杂的分析。
- 它可以与 Hadoop 和现有 Hadoop 数据集成。
- 它提供 Java、Scala 或 Python 的内置 API。
- Spark 提供了内存中数据处理的功能，其速度比 MapReduce 利用的磁盘处理要快得多。
- Spark 可在云和本地与 HDFS、OpenStack 和 Apache Cassandra 配合使用，为企业的大数据操作增加了另一层多功能性。

## 5. Splice Machine

Splice Machine 是一个大数据分析工具。其架构可跨 AWS、Azure 和 Google 等公共云进行移植。

功能：

- 它可以动态地从几个节点扩展到数千个节点，以支持各种规模的应用程序。
- Splice Machine 优化器自动评估分布式 HBase 区域的每个查询。
- 它可以减少管理、更快地部署并降低风险。
- 它能应对快速的流数据，开发、测试和部署机器学习模型。

## 6. Plotly

Plotly 是一种分析工具，可让用户创建图表和仪表板，以进行在线共享。

功能：

- 它可以轻松地将任何数据变成醒目的且信息丰富的图形。
- 它为被审核行业提供有关数据来源的详细信息。
- 它通过免费的社区计划提供不受限制的公共文件托管。

## 7. Azure HDInsight

Azure HDInsight 是云中的 Spark 和 Hadoop 服务。它提供标准和高级两类大数据云产品。它为组织提供了企业规模的集群来运行其大数据工作负载。

功能：

- 它通过行业领先的 SLA 提供可靠的分析。
- 提供企业级安全性和监管。
- 它保护数据资产，并将本地安全性和治理控制扩展到云。
- 对于开发人员和科学家来说，这是一个高生产率的平台。
- 它与领先的生产力应用程序集成在一起。
- 无须购买新硬件或支付其他前期费用，即可在云中部署 Hadoop。

**8. Skytree**

Skytree 是一个大数据分析工具，可让数据科学家更快地构建更准确的模型。它提供了易于使用的准确的预测性机器学习模型。

功能：

- 它具有高度可扩展的算法。
- 它是数据科学家的人工智能。
- 它使数据科学家能够可视化地呈现并了解机器学习决策背后的逻辑。
- 通过 Skytree 轻松使用 GUI 或用 Java 自动编程。
- 具有模型可解释性。
- 它旨在通过数据准备功能解决鲁棒预测问题。
- 它是程序化的并且允许 GUI 访问。

**9. Lumify**

Lumify 被认为是可视化平台、大数据融合和分析工具。它通过一系列分析选项帮助用户发现连接并探索其数据中的关系。

功能：

- 它提供了具有各种自动布局的 2D 和 3D 图形可视化。
- 它提供了图实体之间的链接分析、与地图系统的集成、地理空间分析、多媒体分析以及对一组项目或工作空间的实时协作。
- 它具有针对文本内容、图像和视频的特定处理与界面元素。
- 其空间功能使你可以将工作组织到一组项目或工作区中。
- 它基于成熟的、可扩展的大数据技术。
- 它支持基于云的环境。它可以与 Amazon 的 AWS 完美配合。

**10. Hadoop**

Hadoop 在大数据处理领域久居榜首。它以其大规模数据处理能力而闻名。由于开源的大数据框架可以在本地或云中运行，因此对硬件的要求较低。

功能：

- 使用 HTTP 代理服务器时，它改进了身份验证。
- 它支持 POSIX 样式的文件系统扩展属性。
- 它提供了一个强大的生态系统，完全满足开发人员的分析需求。
- 它带来了数据处理的灵活性。
- 它让数据处理得更快。

# Exercises

**[Ex. 1]** 根据 Text A 回答以下问题。

1. What is ETL?


2. What is the full form of ETL?

3. What is almost essential to the success of a data warehouse project?

4. What happens in the extraction step?

5. How many data extraction methods are there? What are they?

6. Why does the data extracted from source server need to be cleansed, mapped and transformed?

7. What are the types of loading mentioned in the passage?

8. What is MarkLogic? What can it do?

9. What is Oracle? What does it do?

10. What are the best practices of ETL process?

[Ex. 2] 根据 Text B 回答以下问题。

1. What is Apache Storm?

2. What is Talend? What does it do?

3. What is Apache CouchDB? What language is it written in?

4. What is Apache Spark?

5. What are the features of Splice Machine?

6. What are the features of Plotly?

7. What is Azure HDInsight? What does it provide?

8. What is Skytree? What does it offer?

9. What is Lumify considered? What does it do?

10. What is Hadoop?

**[Ex. 3]** 词汇英译中

1. business decision
2. data stream
3. data integrity
4. native code
5. data map
6. proxy server
7. communication protocol
8. analyst
9. sample data
10. duplicate

1. _____
2. _____
3. _____
4. _____
5. _____
6. _____
7. _____
8. _____
9. _____
10. _____

**[Ex. 4]** 词汇中译英

1. 文本文件
2. 过滤
3. 用例
4. 映射
5. 敏捷的，灵活的
6. 恢复；重新获得；找回
7. 擦掉；抹去；清除
8. 样本；样品；抽样调查；取样
9. 间隔尺寸，粒度
10. 验证；证明；证实

1. _____
2. _____
3. _____
4. _____
5. _____
6. _____
7. _____
8. _____
9. _____
10. _____

**[Ex. 5]** 短文翻译

### Data Analytics

Data analytics is a broad term that encompasses many diverse types of data analysis. Any type of information can be subjected to data analytics techniques to get insight that can be used to improve things.

The process involved in data analysis involves several different steps.

1) The first step is to determine the data requirements or how the data is grouped. Data may be separated by age, demographic, income or gender. Data values may be numerical or be divided by category.

2) The second step in data analytics is the process of collecting it. This can be done through a variety of sources such as computers, online sources, cameras, environmental sources or through personnel.

3) Once the data is collected, it must be organized so it can be analyzed. Organization may take place on a spreadsheet or other form of software that can take statistical data.

4) The data is then cleaned up before analysis. This means it is scrubbed and checked to ensure there is no duplication or error, and that it is not incomplete. This step helps correct any errors before it goes on to a data analyst to be analyzed.

Data analytics is broken down into four basic types.

1) Descriptive analytics describes what has happened over a given period of time. Have the number of views gone up? Are sales stronger this month than last?

2) Diagnostic analytics focuses more on why something happened. This involves more diverse data inputs and a bit of hypothesizing. Did the weather affect beer sales? Did that latest marketing campaign impact sales?

3) Predictive analytics is the process of using historical data to analyze past patterns and predict future patterns. What happened to sales the last time we had a hot summer? What will happen if we will have a hot summer this year?

4) Prescriptive analytics suggests a course of action. If the likelihood of a hot summer is measured as an average of these weather models is above 58%, we should add an evening shift to the brewery and rent an additional tank to increase output.

# Reading Material

## Big Data as a Service (BDaaS)

Big Data as a Service[1], which is a combination of big data technologies and cloud computing platforms, helps users to reduce the cost and time for the deployment of big data projects. Moreover, BDaaS enables organizations to manage big data on the cloud and allows all departments to easily access data at any given time.

Essentially[2], BDaaS is any service that involves managing or running big data on the cloud.

**1. The advantages of BDaaS**

There are many advantages to using a BDaaS solution. It makes many of the aspects that managing a big data infrastructure yourself so much easier.

One of the biggest advantages is that it makes managing large quantities of data possible for medium-sized businesses[3]. Not only can it be technically and physically challenging, it can also be expensive. With BDaaS solutions that run in the cloud, companies don't need to stump up[4] cash up front, and operational expenses on hardware can be kept to a minimum. With cloud computing, your infrastructure requirements are fixed at a monthly or annual cost.

However, it's not just about storage and cost. BDaaS solutions sometimes offer in-built[5] solutions

---

1　Big Data as a Service (BDaaS): 大数据即服务
2　essentially [ɪ'senʃəlɪ] *adv.*本质上，根本上
3　medium-sized businesse: 中型企业
4　stump up:付清，掏腰包
5　in-built [,in'bɪlt] *adj.*内置的，固定的，嵌入的

for artificial intelligence and analytics, which means you can accomplish[6] some pretty impressive results without having to have a huge team of data analysts, scientists and architects around you.

## 2. The different models of BDaaS

There are three different BDaaS models. These closely align with the 3 models of cloud infrastructure: IaaS, PaaS and SaaS.

1) Big Data Infrastructure as a Service (IaaS)[7]. Basic data services from a cloud service provider.

2) Big Data Platform as a Service (PaaS)[8]. Offerings of an all-round big data stack like those provided by Amazon S3, EMR or RedShift. This excludes ETL and BI.

3) Big Data Software as a Service (SaaS)[9]. A complete big data stack within a single tool.

## 3. How does the big data Iaas model work

A good example of the IaaS model is Amazon's AWS IaaS architecture, which combines S3 and EC2. Here, S3 acts as a data lake that can store infinite amounts of structured as well as unstructured data. EC2 acts a compute layer that can be used to implement a data service of your choice and connects to the S3 data.

For the data layer you have the option of choosing from among.

● Hadoop. The Hadoop ecosystem can be run on an EC2 instance, giving you complete control.

● NoSQL databases. These include MongoDB or Cassandra.

● Relational databases. These include PostgreSQL or MySQL.

For the compute layer, you can choose from among.

● Self-built ETL scripts[10] that run on EC2 instances.

● Commercial ETL tools that can run on Amazon's infrastructure and use S3.

● Open source[11] processing tools that run on AWS instances, like Kafka.

## 4. How does the big data paas model work

A standard Hadoop cloud-based big data infrastructure on Amazon contains the following.

1) Data Ingestion. Logs file data from any data source.

2) Amazon S3 Data Storage Layer.

● Amazon EMR: A scalable set of instances that run Map/Reduce against the S3 data.

● Amazon RDS: A hosted MySQL database that stores the results from Map/Reduce computations.

3) Analytics and Visualization. Using an in-house[12] BI tool.

A similar set up can be replicated using Microsoft's Azure HDInsight. The data ingestion can be made easier with Azure Data Factory's copy data tool. Apart from that, Azure offers several storage options like data lake storage and Blob storage that you can use to store results

---

6  accomplish [ə'kʌmplɪʃ] v.完成，达成

7  Infrastructure as a Service (IaaS): 基础设施即服务

8  Platform as a Service (PaaS): 平台即服务

9  Software as a Service (SaaS): 软件即服务

10  script [skrɪpt] n.脚本

11  open source: 开源

12  in-house ['ɪnhaʊs] adj.内部的

from the computations.

**5. How does the big data SaaS model work**

A fully hosted big data stack includes everything from data storage to data visualization contains the following.

1) Data Layer. Data needs to be pulled into[13] a basic SQL database. An automated data warehouse does this efficiently.

2) Integration Layer. It pulls the data from the SQL database into a flexible modeling layer.

3) Processing Layer. It prepares the data based on the custom business requirements and logic provided by the user.

4) Analytics and BI Layer. It has fully featured BI abilities which include visualizations, dashboards and charts, etc.

Azure Data Warehouse and AWS Redshift are the popular SaaS options that offer a complete data warehouse solution in the cloud. Their stack integrates all the four layers and is designed to be highly scalable. Google's BigQuery is another contender that's great for generating meaningful insights at an unmatched price-performance.

**6. Conclusion**

The value of big data is not in the data itself, but in the insights that can be drawn after processing it and running it through robust analytics. This can help to guide and define your decision making for the future.

A quick tip with regards to using big data is keep it small at the initial stages. This ensures that the data can be checked for accuracy and the metrics derived from them are right. Once confirmed, you can go ahead with more complex and larger data projects.

---

13　be pulled into: 被拖入，被拉入

# Unit 6

## Text A

### Data Mining

扫码听课文

#### 1. What is data mining

Data mining is the practice of automatically searching large stores of data to discover patterns and trends that go beyond simple analysis. Data mining uses sophisticated mathematical algorithms to segment the data and evaluate the probability of future events. Data mining is also known as Knowledge Discovery in Data (KDD).

(1) Automatic discovery

Data mining is accomplished by building models. A model uses an algorithm to act on a set of data. The notion of automatic discovery refers to the execution of data mining models.

Data mining models can be used to mine the data on which they are built, but most types of models are generalizable to new data. The process of applying a model to new data is known as scoring.

(2) Prediction

Many forms of data mining are predictive. For example, a model might predict income based on education and other demographic factors. Predictions have an associated probability (how likely is this prediction to be true). Prediction probabilities are also known as confidence (how confident can I be of this prediction).

Some forms of predictive data mining generate rules, which are conditions that imply a given outcome. For example, a rule might specify that a person who has a bachelor's degree and lives in a certain neighborhood is likely to have an income greater than the regional average. Rules have an associated support (what percentage of the population satisfies the rule).

(3) Grouping

Other forms of data mining can identify natural groupings in data. For example, a model might identify the segment of the population that has an income within a specified range, that has a good driving record and that leases a new car on a yearly basis.

(4) Actionable information

Data mining can derive actionable information from large volumes of data. For example, a town planner might use a model that predicts income based on demographics to develop a plan for low-income housing. A car leasing agency might a use model that identifies customer segments to design a promotion targeting high-value customers.

## 2. Steps involved in data mining

(1) Business understanding

During data mining, we will understand every aspect of the business objectives and needs. The current situation is assessed by finding the resources, assumptions and other important factors. Accordingly, establishing a good data mining plan will help achieve both business and data mining goals.

(2) Data understanding

Initially, the data is collected from all of the available sources. Then we choose the best data set from where we can extract the data which could be more beneficial.

(3) Data preparation

Once the data set is identified, it is selected, cleaned, constructed and formatted in the desired form.

(4) Data modelling

It is a process of remodeling the given data according to the requirement of the user. One or more models could be created on the prepared data set and finally, the models need to be assessed carefully by stakeholders to make sure that created models meet business initiatives.

(5) Evaluation

This is one of the most necessary process in data mining. It includes going through every aspect of the process so as to check for any possible fault or data leakage in the process. Also, new business requirements could be raised due to the new patterns discovered.

(6) Deployment

It means to simply present the knowledge in such a way that the stakeholders can use it when they want it. For example, it is found that there are fewer international calls on Wednesday. When this information is presented to the stakeholders, they in turn can use this information to their advantage and increase their profits.

## 3. Types of data mining

(1) Smoothing

This particular method of data mining technique comes under the genre of preparing the data. The main intent of this technique is to remove noise from the data. Here algorithms like simple exponential, the moving average are used to remove the noise. During exploratory analysis, this technique is very handy to visualize trends/sentiments.

(2) Aggregation

As the term suggests, a group of data is aggregated to achieve more information. This technique is employed to give an overview of business objectives and can be performed manually or using specialized software. This technique is generally employed on big data, as big data don't provide the required information as a whole.

(3) Generalization

Again, as the name suggests, this technique is employed to generalize data as a whole. This is different from aggregation in that the data during generalization is not grouped together to achieve more information but in turn, the entire data set is generalized. This will enable a data science model to adapt to newer data points.

(4) Normalization

In this technique, special care is employed to data points so as to bring them into the same scale for analysis. For example, the age and salary of a person fall in different measurement scales, hence plotting them on a graph won't help us attain any useful info about the trends present as a collective feature. Using normalization, we can bring them into an equal scale so that apple to apple comparison can be performed.

(5) Attribute/Feature selection

In this technique, we employ methods to perform a selection of features so that the model used to train the data sets can imply value to predict the data it has not seen. This is very analogous to choosing the right outfit from a wardrobe full of clothes to fit oneself right for the event. Non-relevant features can negatively impact model performance, let alone improving performance.

(6) Classification

In this technique of data mining we deal with groups known as "classes". We employ the features selected collectively to groups/categories. For example, in a shop, if we have to evaluate whether a person will buy a product or not, there are "n" number of features we can collectively use to get a result of True/False.

(7) Pattern tracking

This is one of the basic techniques employed in data mining to get information about trends/patterns which might be exhibited by the data points. For example, we can determine a trend of more sales during a weekend or holiday time rather than on weekdays or working days.

(8) Outlier analysis or anomaly detection

This technique is used for finding or analyzing outliers or anomalies. Outliers or anomalies are not negative data points, they are just something that stands out from the general trend of the entire data set. On identifying the outliers, we can either remove them completely from the data set, which occurs when the preparation of data is done. Or else this technique is extensively used in model data sets to predict outliers as well.

(9) Clustering

This technique is pretty much similar to classification, but the only difference is we don't know the group in which data points will fall. This method is typically used in grouping people to target similar product recommendations.

(10) Regression

This technique is used to predict the likelihood of a feature with the presence of other features. For example, we can formulate the likelihood of the price of an item with respect to demand, competition and a few other features.

(11) Neural network

This technique is based on the principle of how biological neurons work. Similar to what neurons in the human body do, the neurons in a neural network in data mining also work act as the processing unit and connecting another neuron to pass on the information along the chain.

(12) Association

In this method of data mining, the relation between different features are determined and in turn, used either to find hidden patterns or to perform related analysis as per business requirement. For example, using the association we can find features correlated to each other, remove a feature in order to remove some redundant features and improve processing power/time.

To conclude, there are different requirements one should keep in mind while data mining is performed. One needs to be very careful of what the output is expected to be so that corresponding techniques can be used to achieve the goal.

# ✑ New Words

| discover | [dɪ'skʌvə] | vt.发现；获得知识 |
|---|---|---|
| trend | [trend] | n.趋势；走向 |
| mathematical | [ˌmæθə'mætɪkl] | adj.数学的 |
| segment | ['segmənt] | n.部分，段落 |
| | | v.分段，分割；划分 |
| probability | [ˌprɒbə'bɪləti] | n.可能性；概率 |
| execution | [ˌeksɪ'kju:ʃn] | n.实行，履行，执行 |
| generalizable | ['dʒenərəˌlaɪzəbl] | adj.可泛化的，可推广的 |
| prediction | [prɪ'dɪkʃn] | n.预测，预报；预言 |
| demographic | [ˌdemə'græfik] | adj.人口统计学的 |
| factor | ['fæktə] | n.因素 |
| confidence | ['kɒnfɪdəns] | n.信心，信任 |
| confident | ['kɒnfɪdənt] | adj.确信的，深信的 |
| condition | [kən'dɪʃn] | n.状态；环境 |
| | | vt.制约，限制；使习惯于，使适应 |
| imply | [ɪm'plaɪ] | v.暗示，意味，隐含 |
| neighborhood | ['neɪbəhʊd] | n.地区；某地区的人 |
| regional | ['ri:dʒənl] | adj.地区的，区域的 |
| population | [ˌpɒpju'leɪʃn] | n.人口 |
| satisfy | ['sætɪsfaɪ] | vt.使确信；符合，达到（要求、规定、标准等） |
| | | vi.使满足或足够 |
| record | ['rekɔ:d] | n.记录；档案 |
| derive | [dɪ'raɪv] | v.得到，导出；源于 |
| promotion | [prə'məʊʃn] | n.促进；提升，升级；（商品等的）推广 |
| aspect | ['æspekt] | n.方面；层面；样子；外观 |
| situation | [ˌsɪtʃu'eɪʃn] | n.情况；形势，处境；位置 |

| | | |
|---|---|---|
| assumption | [ə'sʌmpʃn] | n.假定，假设 |
| beneficial | [,benɪ'fɪʃl] | adj.有利的，有益的 |
| preparation | [,prepə'reɪʃn] | n.准备，预备 |
| construct | [kən'strʌkt] | vt.构建，构造 |
| remodeling | [ri:'mɒdlɪŋ] | v.重构，改造 |
| requirement | [rɪ'kwaɪəmənt] | n.需求，要求；必要条件 |
| initiative | [ɪ'nɪʃətɪv] | n.主动性；主动权；倡议 |
| leakage | ['li:kɪdʒ] | n.漏出；泄露 |
| deployment | [dɪ'plɔɪmənt] | n.部署；调度 |
| profit | ['prɒfit] | n.利润；红利 |
| | | vi.有益；获利 |
| smoothing | ['smu:ðɪŋ] | v.（使）光滑，（使）平坦 |
| genre | ['ʒɒnrə] | n.类型，种类 |
| exponential | [,ekspə'nenʃl] | adj.指数的，幂数的 |
| exploratory | [ɪk'splɒrətri] | adj.探索的 |
| handy | ['hændi] | adj.方便的；手边的；便于使用的 |
| sentiment | ['sentɪmənt] | n.感情，情绪；意见，观点 |
| achieve | [ə'tʃi:v] | vt.取得，获得；实现 |
| generalization | [,dʒenrəlaɪ'zeɪʃn] | n.泛化，概括 |
| generalize | ['dʒenrəlaɪz] | v.泛化；推广 |
| normalization | [,nɔ:məlaɪ'zeɪʃn] | n.归一化 |
| attain | [ə'teɪn] | v.获得；到达 |
| train | [treɪn] | v.训练 |
| analogous | [ə'næləgəs] | adj.相似的，可比拟的 |
| collectively | [kə'lektɪvli] | adv.全体地，共同地 |
| outlier | ['aʊtlaɪə] | n.离群值；异常值 |
| anomaly | [ə'nɒməli] | n.异常，反常 |
| clustering | ['klʌstərɪŋ] | n.聚类 |
| regression | [rɪ'greʃn] | n.回归 |
| likelihood | ['laɪklɪhʊd] | n.可能，可能性；[数]似然，似真 |
| formulate | ['fɔ:mjuleɪt] | vt.构想出，规划；用公式表示 |
| neuron | ['njʊərɒn] | n.神经元；神经细胞 |
| chain | [tʃeɪn] | n.链路，链条 |
| association | [ə,səʊʃi'eɪʃn] | n.关联；联合，联系 |
| corresponding | [,kɒrə'spɒndɪŋ] | adj.相当的，对应的；符合的 |

# 🔖 Phrases

| | |
|---|---|
| build models | 建立模型，构建模型 |
| town planner | 城市规划者 |
| high-value customer | 高价值的客户 |
| data leakage | 数据泄露 |
| be exhibited by | 用……展示 |
| stand out | 突出；出色；更为重要；引人注目；显眼 |
| neural network | 神经网络 |
| processing unit | 处理单元 |
| hidden pattern | 隐藏模式，隐含模式 |

# 🔖 Abbreviations

KDD (Knowledge Discovery in Data)　　数据知识发现

## Reference Translation

# 数 据 挖 掘

## 1. 什么是数据挖掘

数据挖掘是一种实践活动，它可以对大量数据进行自动搜索，目的在于发现用简单分析不能找到的模式和趋势。数据挖掘使用复杂的数学算法来分割数据并评估未来事件的可能性。数据挖掘也称为数据知识发现（KDD）。

（1）自动发现

数据挖掘是通过构建模型来完成的，模型将算法用于一组数据。自动发现的概念是指执行数据挖掘模型。

数据挖掘模型可用于挖掘构建它们的数据，但是大多数类型的模型均可泛化到新数据。将模型应用于新数据的过程被称为评分。

（2）预测

许多形式的数据挖掘都是可预测的。例如，模型可以根据教育程度和其他人口统计因素来预测收入。预测具有关联的概率（此预测为真的可能性有多大）。预测概率也称为置信度（对这个预测有多少信心）。

某些形式的预测性数据挖掘会生成规则，它们是显示给定结果的条件。例如，一条规则可能会指定拥有学士学位并且居住在某个社区中的人的收入可能会高于该区域的平均水平。规则具有相关的支持（满足规则的人口百分比是多少）。

（3）分组

其他形式的数据挖掘可以识别数据中的自然分组。例如，模型可能会确定收入在指定范围内，具有良好驾驶记录并每年租赁新车的人群。

（4）可操作信息

数据挖掘可以从大量数据中获取可操作信息。例如，城镇规划人员可能会使用基于人口统计数字预测收入的模型来制订低收入住房计划。汽车租赁公司可能会使用一种识别客户细分的模型，以设计针对高价值客户的促销活动。

## 2. 数据挖掘的步骤

（1）理解业务

在数据挖掘的过程中，我们将了解业务目标和需求的各个方面。通过查找资源、假设和其他重要因素来评估当前状况。因此，制订良好的数据挖掘计划将有助于实现业务和数据挖掘目标。

（2）理解数据

最初，从所有可用来源收集数据。然后，选择最佳数据集，从中提取可能更有益的数据。

（3）准备数据

识别出数据集后，将以所需的形式对其进行选择、清理、构建和格式化。

（4）数据建模

这是根据用户要求重新构建给定数据的过程。可以根据准备好的数据集创建一个或多个模型，最后，利益相关者需要仔细评估模型，以确保创建的模型符合业务计划。

（5）评估

这是数据挖掘中最必要的过程之一。它包括遍历过程的各个方面，以检查过程中是否存在任何可能的故障或数据泄露。此外，由于发现了新的模式，可能会提出新的业务需求。

（6）部署

它意味着只呈现知识，以使利益相关者可以在需要时使用它。例如，发现星期三的国际电话较少。当将这些信息提供给利益相关者时，他们反过来可以利用这些信息使自己受益，并增加他们的利润。

## 3. 数据挖掘的类型

（1）平滑

数据挖掘技术的这种特定方法属于准备数据的范畴。该技术的主要目的是消除数据中的噪声。这里使用简单指数、移动平均值之类的算法来消除噪声。在探索性分析中，此技术非常便于可视化趋势/情感。

（2）聚合

顾名思义，此技术就是聚合一组数据以获得更多信息。该技术用于总览业务目标，可以手动执行，也可以使用专用软件执行。这种技术通常用于大数据，因为大数据不能提供整体所需的信息。

（3）泛化

再次，顾名思义，这种技术被用来整体上泛化数据。这与聚合不同，因为泛化过程中的数据不是为了获取更多信息而组合在一起，而是对整个数据集进行了泛化。这将使数据科学模型能够适应更新的数据点。

（4）归一化

在此技术中，对数据点采取了特殊的措施，以便将它们置于相同的范围内进行分析。例如，一个人的年龄和薪水属于不同的度量标准，因此将其绘制在一个图表上并不会帮助我们获得有关作为整体特征呈现的趋势的任何有用信息。使用归一化，我们可以将它们放到同等类别，以便可以进行逐一比较。

（5）属性/功能选择

在这项技术中，我们采用一些方法来执行特征选择，以便用于训练数据集的模型可以表明价值以预测未看到的数据。这非常类似于从装满衣服的衣柜中选择与自己的活动相匹配的衣服。不相关的功能可能会对模型性能产生负面影响，更不用说提高性能了。

（6）分类

在这种数据挖掘技术中，我们处理称为"类"的组，将选择的共同特征应用于组/类别。例如，在一家商店中，如果我们必须评估一个人是否会购买某种产品，那么我们可以集体使用"n"个特征来获得对/错的结果。

（7）模式跟踪

这是数据挖掘中所用的基本技术之一，用来获取有关可能由数据点显示的趋势/模式信息。例如，我们可以确定在周末或节假日销售有增加的趋势，而不是在平日或工作日。

（8）离群分析或异常检测

此技术用于查找或分析离群值或异常值。离群值或异常值不是负数据点，它们只是与整个数据集的总体趋势不同。在识别异常值时，我们可以将它们从数据集中完全删除，这在完成数据准备时会发生。或者把该技术广泛用于模型数据集中以预测离群值。

（9）聚类

该技术与分类非常相似，但唯一的区别是我们不知道数据点所属的组。这种方法通常用于将人们分组，以获得类似的产品推荐。

（10）回归

该技术用于预测存在其他特征时某个特征存在的可能性。例如，我们可以根据需求、竞争和其他一些特征来规划商品可能的价格。

（11）神经网络

该技术基于生物神经元的工作原理。与人体神经元的原理相似，数据挖掘工作中神经网络的神经元也充当处理单元，并连接另一个神经元以便沿着链路传递信息。

（12）关联

在这种数据挖掘方法中，确定不同特征之间的关系，然后根据业务需求将其用于查找隐藏模式或进行相关分析。例如，使用关联，我们可以找到相互关联的特征，删除某个特征以便去除一些冗余特征并提高处理能力/时间。

总之，在执行数据挖掘时应牢记不同的要求。需要非常小心地对待预期的输出，以便可以使用相应的技术来实现目标。

# Text B

扫码听课文

## Top 10 Most Common Data Mining Algorithms

**1. C4.5 algorithm**

C4.5 is one of the top data mining algorithms and was developed by Ross Quinlan. C4.5 is used to generate a classifier in the form of a decision tree from a set of data that has already been classified. Classifier here refers to a data mining tool that takes data we need to classify and tries to predict the class of new data.

Every data point will have its own attributes. The decision tree created by C4.5 poses a question about the value of an attribute and depending on those values, the new data gets classified. The training dataset is labelled, making C4.5 a supervised learning algorithm. Decision trees are always easy to interpret and explain, which makes C4.5 fast and popular compared to other data mining algorithms.

## 2. K-means algorithm

K-means is one of the most common clustering algorithms. It works by creating a K number of groups from a set of objects based on the similarity between objects. It may not be guaranteed that group members will be exactly similar, but group members will be more similar as compared to non-group members. As per standard implementations, K-means is an unsupervised learning algorithm as it learns the cluster on its own without any external information.

## 3. Support vector machine

In terms of tasks, Support Vector Machine (SVM) works similar to C4.5 algorithm except that SVM doesn't use any decision trees at all. SVM learns the datasets and defines a hyperplane to classify data into two classes. A hyperplane is an equation for a line that looks something like "y = mx + b". SVM exaggerates to project your data to higher dimensions. Once projected, SVM has defined the best hyperplane to separate the data into the two classes.

## 4. Apriori algorithm

Apriori algorithm works by learning association rules. Association rules are a data mining technique that is used for learning correlations between variables in a database. Once the association rules are learned, it is applied to a database containing a large number of transactions. Apriori algorithm is used for discovering interesting patterns and mutual relationships and hence is treated as an unsupervised learning approach. Though the algorithm is highly efficient, it consumes a lot of memory, utilizes a lot of disk space and takes a lot of time.

## 5. Expectation Maximization algorithm

Expectation Maximization (EM) is used as a clustering algorithm, just like the K-means algorithm for knowledge discovery. EM algorithm works in iterations to optimize the chances of seeing observed data. Next, it estimates the parameters of the statistical model with unobserved variables, thereby generating some observed data. EM algorithm is also unsupervised learning algorithm since we are using it without providing any labelled class information.

## 6. PageRank algorithm

PageRank is commonly used by search engines like Google. It is a link analysis algorithm that determines the relative importance of an object linked within a network of objects. Link analysis is a type of network analysis that explores the associations among objects. Google search uses this algorithm by understanding the backlinks between web pages.

PageRank is one of the methods Google uses to determine the relative importance of a web page and rank it higher on Google search engine. The PageRank trademark is proprietary of Google and the PageRank algorithm is patented by Stanford University. PageRank is treated as an unsupervised learning approach as it determines the relative importance just by considering the links and doesn't require any other inputs.

**7. Adaboost algorithm**

Adaboost is a boosting algorithm used to construct a classifier. A classifier is a data mining tool that takes data and predicts the class of the data based on inputs. Boosting algorithm is an ensemble learning algorithm which runs multiple learning algorithms and combines them.

Boosting algorithms take a group of weak learners and combine them to make a single strong learner. A weak learner classifies data with less accuracy. The best example of a weak algorithm is the decision stump algorithm which is basically a one-step decision tree. Adaboost is perfect supervised learning as it works in iterations and in each iteration, it trains the weaker learners with the labelled dataset. Adaboost is a simple and pretty straightforward algorithm.

After the user specifies the number of rounds, each successive Adaboost iteration redefines the weights for each of the best learners. This makes Adaboost a great way to auto-tune a classifier. Adaboost is flexible, versatile and elegant as it can incorporate most learning algorithms and can take on a large variety of data.

**8. KNN algorithm**

KNN is a lazy learning algorithm used as a classification algorithm. A lazy learner will not do anything during the training process except for storing the training data. Lazy learners start classifying only when new unlabeled data is given as an input. Besides, C4.5, SVN and Adaboost are eager learners that start to build the classification model during training itself. Since KNN is given a labelled training dataset, it is treated as a supervised learning algorithm.

**9. Naive Bayes algorithm**

Naive Bayes is not a single algorithm though it can be seen as a single algorithm. Naive Bayes is a bunch of classification algorithms put together. The assumption used by the family of algorithms is that every feature of the data being classified is independent of all other features that are given in the class. Naive Bayes is provided with a labelled training dataset to construct the tables. So it is treated as a supervised learning algorithm.

**10. CART algorithm**

CART stands for classification and regression trees. It is a decision tree learning algorithm that gives either regression or classification trees as an output. In CART, the decision tree nodes will have precisely 2 branches. Just like C4.5, CART is also a classifier. The regression or classification tree model is constructed by using labelled training dataset provided by the user. Hence it is treated as a supervised learning algorithm.

 New Words

| generate | ['dʒenəreɪt] | vt.形成，造成；引起 |

| classifier | ['klæsɪfaɪə] | n.分类器，分类者 |
| interpret | [ɪn'tɜ:prət] | v.解释 |
| similarity | [ˌsɪmə'lærəti] | n.类似；相似点 |
| implementation | [ˌɪmplɪmen'teɪʃn] | n.实施 |
| cluster | ['klʌstə] | n.丛；簇，串；群<br>vi.丛生；群聚<br>vt.使聚集 |
| hyperplane | ['haɪpəˌpleɪn] | n.超平面 |
| equation | [ɪ'kweɪʒn] | n.方程式；等式 |
| exaggerate | [ɪg'zædʒəreɪt] | v.夸张；夸大 |
| correlation | [ˌkɒrə'leɪʃn] | n.相互关系；相关性 |
| utilize | ['ju:təlaɪz] | vt.利用，使用 |
| iteration | [ˌɪtə'reɪʃn] | n.迭代；循环 |
| estimate | ['estɪmət]<br>['estɪmeɪt] | n.估价，估算<br>v.估价，估算 |
| backlink | ['bæklɪŋk] | n.反向链接 |
| proprietary | [prə'praɪətri] | adj.专有的，专利的 |
| ensemble | [ɒn'sɒmbl] | n.集成；全体 |
| redefine | [ˌri:dɪ'faɪn] | v.重新定义，再定义 |

## Phrases

| in the form of | 用……的形式 |
| decision tree | 决策树 |
| clustering algorithm | 聚类算法 |
| Apriori algorithm | Apriori 算法 |
| association rule | 关联规则 |
| mutual relationship | 相互关系 |
| disk space | 磁盘空间 |
| search engine | 搜索引擎 |
| link analysis algorithm | 链接分析算法，链路分析算法 |
| boosting algorithm | 提升算法 |
| weak learner | 弱学习器 |
| strong learner | 强学习器 |
| decision stump algorithm | 决策树桩算法 |
| lazy learning algorithm | 惰性学习算法，消极学习算法 |
| unlabeled data | 未标记的数据 |
| a bunch of | 一束；一群；一堆 |
| be constructed by | 由……构建 |

## ✎ Abbreviations

SVM (Support Vector Machine)                        支持向量机

EM (Expectation Maximization)                        期望最大化

KNN (K-Nearest Neighbor)                             K 近邻算法

CART (Classification and Regression Trees)          分类与回归树

## Reference Translation

# 十大最常见的数据挖掘算法

## 1. C4.5 算法

C4.5 是顶级数据挖掘算法之一，由 Ross Quinlan 开发。C4.5 用于根据已分类的一组数据，以决策树的形式生成分类器。这里的分类器是指一种数据挖掘工具，该工具获取我们需要分类的数据并尝试预测新数据的类别。

每个数据点将具有自己的属性。由 C4.5 创建的决策树提出了有关属性值的问题，并根据这些值对新数据进行分类。其提供带标签的训练数据集，C4.5 被视为监督学习算法。与其他数据挖掘算法相比，决策树始终易于解释和说明，这使 C4.5 快速且流行。

## 2. K-均值算法

K-均值是最常见的聚类算法之一。它的工作方式是根据对象之间的相似性，从一组对象中创建 K 个组。也许不能保证组成员完全相似，但是与非组成员相比，组成员将更加相似。按照标准实现，K-均值是一种无监督学习算法，因为它无须任何外部信息即可自行学习聚类。

## 3. 支持向量机

就任务而言，支持向量机（SVM）的工作方式类似于 C4.5 算法，但支持向量机根本不使用任何决策树。支持向量机学习数据集并定义一个超平面以将数据分为两类。超平面是一条线的方程，看起来像 "y = mx + b"。支持向量机会把数据扩展映射到更高的维度。一旦映射，支持向量机便定义了最佳的超平面，将数据分为两个类别。

## 4. Apriori 算法

Apriori 算法通过了解关联规则来工作。关联规则是一种数据挖掘技术，用于了解数据库中变量之间的相关性。一旦了解了关联规则，就将其应用于包含大量事务的数据库。Apriori 算法用于发现有趣的模式和相互关系，因此被视为无监督学习算法。尽管该算法是高效的，但它会消耗大量内存、占用大量磁盘空间并花费大量时间。

## 5. 期望最大化算法

期望最大化（EM）用作聚类算法，就像用于知识发现的 K-均值算法一样。EM 算法以迭代方式工作，以更好地查看观测数据。接下来，它估计带有未观察到变量的统计模型的参数，从而生成一些观察到的数据。EM 算法也是无监督学习算法，因为我们在不提供任何标记的类信息的情况下使用它。

## 6. PageRank 算法

PageRank 通常被谷歌等搜索引擎使用。它是一种链接分析算法，可确定对象网络中链接对象的相对重要性。链接分析是一种探索对象之间关联的网络分析。谷歌搜索通过了解网页之间的反向链接来使用此算法。

PageRank 是谷歌用来确定网页的相对重要性并使其在谷歌搜索引擎上获得更高排名的方法之一。PageRank 商标是谷歌的专有商标，PageRank 算法是斯坦福大学的专利。PageRank 被视为一种无监督学习方法，因为它仅通过考虑链接即可确定相对重要性，而不需要任何其他输入。

## 7. Adaboost 算法

Adaboost 是一种用于构建分类器的提升算法。分类器是一种数据挖掘工具，可获取数据并根据输入预测数据的类别。提升算法是一种集成学习算法，可运行多种学习算法并将其组合。

提升算法吸收一组弱学习器并将它们组合成一个单一的强学习器。弱学习器对数据进行分类的准确性较低。弱算法的最佳示例是决策树桩算法，它基本上是一个单步决策树。Adaboost 是完美的监督学习，因为它可以以迭代方式工作，并且在每次迭代中，都使用标记的数据集训练较弱的学习器。Adaboost 是一种简单且非常直接的算法。

在用户指定轮数之后，每次连续的 Adaboost 迭代都会为每个最佳学习器重新定义权重。这使 Adaboost 成为自动调整分类器的绝佳方式。 Adaboost 具有灵活性、多功能性和简洁性，因为它可以合并大多数学习算法并可以处理大量数据。

## 8. KNN 算法

KNN 是一种用作分类算法的消极学习算法。消极学习器在训练过程中除了存储训练数据外不会做任何事情。消极学习器仅在新的未标记数据作为输入时才开始分类。另一方面，C4.5、SVN 和 Adaboost 是积极学习器，它们在训练过程中就开始建立分类模型。由于为 KNN 提供了带标签的训练数据集，因此将其视为监督学习算法。

## 9. 朴素贝叶斯算法

朴素贝叶斯不是单个算法，尽管可以将其视为单个算法。朴素贝叶斯是一组分类算法。该算法家族使用的假设是，要分类的数据的每个特征都独立于该类中给出的所有其他特征。朴素贝叶斯用提供给自己的带有标签的训练数据集来构造表格。因此，它被视为监督学习算法。

## 10. CART 算法

CART 代表分类树和回归树。它是一种决策树学习算法，可将回归树或分类树作为输出。在 CART 中，决策树节点将恰好具有两个分支。就像 C4.5 一样，CART 也是一个分类器。回归或分类树模型通过使用用户提供的带标签的训练数据集来构建。因此，它被视为监督学习算法。

# Exercises

**[Ex. 1]** 根据 Text A 填空。

1. Data mining is the practice of automatically searching _____ to discover _____

that go beyond simple analysis. Data mining is also known as _____.

2. Data mining is accomplished by _____. A model uses _____ to act on a set of data. The notion of automatic discovery refers to _____.

3. Predictions have _____. Prediction probabilities are also known as _____. Some forms of predictive data mining _____, which are conditions that imply _____.

4. Data modelling is a process of _____ according to the requirement of _____. One or more models could be created on _____ and finally, the models need to be assessed carefully by _____ to make sure that created models _____.

5. Evaluation is one of _____ in data mining. It includes going through _____ so as to check for _____or data leakage in the process. Also, new business requirements could be raised due to _____.

6. Deployment means to _____ in such a way that the stakeholders can use it _____.

7. Smoothing comes under the genre of _____. The main intent of this technique is _____ from the data. Here algorithms like_____, _____ are used to remove the noise.

8. In normalization, special care is employed to _____ so as to bring them into _____ for analysis.

9. Pattern tracking is one of the basic techniques employed in data mining to _____ about trends/patterns which might be exhibited by _____. Clustering is typically used in _____ to target _____.

10. Regression is used to _____ of a feature with _____. In association, _____ are determined and in turn, used either to _____ or to _____ as per business requirement.

**[Ex. 2]** 根据 Text B 回答以下问题。

1. What is C4.5? What is it used to do?

2. What is K-means? How does it work?

3. How does Apriori algorithm work? What are association rules?

4. Why is the EM algorithm unsupervised learning?

5. What is PageRank?

6. What does Google use PageRank to do?

7. What is Boosting algorithm? What do Boosting algorithms do?

8. What is the difference between lazy learners and eager learners?

9. What is the assumption used by the family of algorithms? What is Naive Bayes provided with?

10. What does CART stand for? What is it?

## [Ex. 3] 词汇英译中

1. build models      1. _____

2. anomaly      2. _____

3. data leakage      3. _____

4. clustering      4. _____

5. hidden pattern      5. _____

6. construct      6. _____

7. neural network      7. _____

8. deployment      8. _____

9. processing unit      9. _____

10. factor      10. _____

## [Ex. 4] 词汇中译英

1. Apriori 算法      1. _____

2. 泛化，概括      2. _____

3. 关联规则      3. _____

4. 神经元；神经细胞      4. _____

5. 聚类算法      5. _____

6. 可能性；概率      6. _____

7. 决策树      7. _____

8. 回归      8. _____

9. 惰性学习算法，消极学习算法      9. _____

10. 分类器，分类者      10. _____

**[Ex. 5]** 短文翻译

## Types of Data Sources in Data Mining

### 1. Flat files

Flat files is defined as data files in text form or binary form with a structure that can be easily extracted by data mining algorithms. Data stored in flat files have no relationship or path among themselves, like if a relational database is stored on a flat file, then there will be no relations between the tables. Flat files are represented by data dictionary,such as CSV file.

### 2. Relational databases

A relational database is defined as the collection of data organized in tables with rows and columns. Physical schema in relational databases is a schema which defines the structure of tables. Logical schema in relational databases is a schema which defines the relationship among tables.

### 3. Data warehouses

A data warehouse is defined as the collection of data integrated from multiple sources. There are three types of data warehouse: enterprise data warehouse, data mart and virtual warehouse. Two approaches can be used to update data in data warehouse: query-driven approach and update-driven approach.

### 4. Transactional databases

Transactional databases is a collection of data organized by timestamps, date, etc. to represent transaction in databases. This type of database has the capability to roll back or undo its operation when a transaction is not completed or committed. It is a highly flexible system where users can modify information without changing any sensitive information.

### 5. Multimedia databases

Multimedia databases consist of audio, video, images and text media. They can be stored on object-oriented databases. They are used to store complex information in a pre-specified formats.

### 6. WWW

WWW refers to World Wide Web. It is a collection of documents and resources like audio, video, text, etc. which are identified by Uniform Resource Locators (URL) through web browsers, linked by HTML pages, and accessible via the Internet network. It is the most heterogeneous repository as it collects data from multiple resources. It is dynamic in nature as volume of data is continuously increasing and changing.

# Reading Material

## Applications of Data Mining

### 1. In healthcare

Data mining holds great potential[1] in the healthcare sector. Data and analytics can be used to

---

1    potential [pə'tenʃl] *adj*.潜在的 *n*.潜力

identify best practices as well as provide cost-effective solutions. The data mining approach includes multi-dimensional databases, statistics, machine learning, data visualization and soft computing that can have massive applications in the industry. It can help predict the volume of patients in every category, improve processes to ensure that patients receive appropriate care without delays or setbacks[2], detect fraud and abuse[3] for insurance purposes and many more.

## 2. In banking and finance

With the advent of digitization, the banking sector is handling and managing enormous amounts of data and transaction information. Data mining applications in banking can easily be the appropriate solution with its capability of identifying patterns, casualties, market risks and other correlations that are crucial for managers to be aware of. Despite the volumes of data, results can be generated almost instantly for the managers to make sense of without much effort.

Bank officials and employees will find the use of all the information to improve and optimize segmentation[4], targeting, acquisition, management and retainment of profitable customers. Data mining can also help banks quickly determine potential defaulters and accordingly make decisions for the issuance of credit cards, loans, etc.

## 3. In customer segmentation

Traditional methods for market research have always been used for the segmentation of customers, but data mining can be more granular, and it can increase effectiveness. It aids in the segmentation process in a more precise manner and helps in the tailoring of customer requirements. Data mining can help identify a customer segment based on vulnerability[5] and enhance customer satisfaction through decisions based on the insights generated.

## 4. In education

Educational Data Mining (EDM) is the latest emerging field that aims to establish techniques to uncover knowledge based on data originating from educational environments. EDM aims to predict students' potential learning behavior, explore the impact of educational support and advance scientific knowledge about learning.

Data mining can be implemented by educational institutions[6] to make sound decisions and predict the achievement levels of students to pay more attention to the content and techniques of teaching. These teaching techniques can be developed by observing the studying and behavioral patterns of the students.

## 5. In market basket analysis

Market basket analysis is one of the key modeling techniques adopted by retailers[7] to identify

---

2  setback ['setbæk] n.挫折；阻碍

3  abuse [ə'bju:s] n.滥用

4  segmentation [ˌsegmen'teɪʃn] n.细分，切分

5  vulnerability [ˌvʌlnərə'bɪləti] n.弱点，攻击

6  educational institution：教育机构

7  retailer ['ri:teɪlə] n.零售商，零售店

the relations between certain groups of items. To put it simply, it looks for combinations of items that are frequently ordered together.

Market basket analysis allows the understanding of the purchase behavior of buyers. This information can come really handy for the retailers who want to know the buyers' requirements and accordingly arrange their stores' layout. Through the process of differential analysis, comparison can be seamlessly carried out between different stores and customers from different demographic groups.

## 6. In fraud detection

Detection of fraud through traditional methods is very complex and time-consuming. This is where data mining comes in to generate valid information and meaningful insights. Ideally, a robust fraud detection system will thoroughly protect user information. Through means of a collection of sample records, a model is built to identify and classify this information as either fraudulent or non-fraudulent.

## 7. In CRM

Customer Relationship Management (CRM) implements customer-focused strategies to acquire customers, improve customer loyalty and retain them. Data mining plays a significant[8] part in business by ensuring healthy relationships with customers. This is done primarily through the data mining technology that collects relevant data and information for analysis. More effective solutions can be generated from the insights received.

## 8. In manufacturing engineering

Often, the manufacturing process can be quite complex and in those situations, valuable and reliable information and knowledge can be a huge asset. This is where data mining tools can be of use. They help identify patterns and trends, and extract the relationships between product portfolio[9], product architecture and customer requirements in system-level designing. Data mining can also aid in the prediction of the product development cost, wear and tear of production assets, span time, dependencies, etc. Manufacturers can anticipate[10] maintenance, which can successfully reduce the downtime.

## 9. In research analysis

Data mining is instrumental[11] in data cleaning, data pre-processing and database integration, which makes it ideal for researchers. Data mining can help identify the correlation between activities or co-occurring sequences that can bring about change in the research. Data mining, in conjunction with data visualization and visual data mining, can offer clarity[12] in data and research.

## 10. In criminal investigation

Crime analysis is concerned with exploring and identifying crime characteristics and studying

---

8   significant [sɪɡ'nɪfikənt] *adj*.重要的，显著的

9   product portfolio：产品组合

10   anticipate [æn'tɪsɪpeɪt] *vt*.预期，预见，预料

11   instrumental [ˌɪnstrə'mentl] *adj*.有帮助的，起作用的

12   clarity ['klærəti] *n*.清楚，明确；思路清晰

their relationships with criminals. A very high volume of datasets exists for criminology[13], due to which it can be very complicated. Evidently, data mining has found extensive applications in this field as an appropriate tool. Conversions of all text-based crime reports into word processing files are possible. These files can further be utilized in crime-matching processes.

Apart from the 10 listed applications of data mining, there are more sectors that make use of it extensively: marketing, intrusion detection, lie detection, corporate surveillance, bioinformatics[14], e-commerce, retail, service providers, insurance, communications and many more.

---

13    criminology [ˌkrɪmɪˈnɒlədʒi] n.犯罪学，刑事学
14    bioinformatics [baɪəuˌɪnfəˈmætɪks] n.生物信息学；生物资讯

# Unit 7

## Text A

## Five Programming Languages for Big Data

扫码听课文

One of the most important decisions that big data professionals have to make, especially the ones who are new to the scene or are just starting out, is choosing the best programming languages for big data manipulation and analysis. Understanding the big data problem and framing the architecture to solve it is not quite enough these days. The execution needs to be perfect as well, and choosing the right language goes a long way. The following are the top five programming languages for big data.

### 1. Python

Python has been declared as one of the fastest growing programming languages according to the recently held Stack Overflow Developer Survey. Its general-purpose nature means it can be used across a broad spectrum of use-cases, and big data programming is one major area of application.

Many libraries for data analysis and manipulation which are increasingly being used in a big data framework to clean and manipulate large chunks of data, such as Pandas, NumPy and SciPy are all Python-based. Not just that, most popular machine learning and deep learning frameworks such as Scikit-Learn, TensorFlow and many more, are also written in Python and are finding increasing application within the big data ecosystem.

One drawback of using Python, and a reason why it is not a first-class citizen when it comes to big data programming yet, is that it's slow. Although very easy to use, big data professionals have found systems built with languages such as Java or Scala are faster and more robust to use than the systems built with Python.

However, Python makes up for this limitation with other qualities. As Python is primarily a scripting language, interactive coding and development of analytical solutions for big data becomes very easy. Python can integrate effortlessly with the existing big data frameworks such as Apache Hadoop and Apache Spark, allowing you to perform predictive analytics at scale without any problem.

Why do we use Python for big data?

- It is general-purpose.
- It has rich libraries for data analysis and machine learning.
- It is easy to use.

- It supports iterative development.
- It has rich integration with big data tools.
- It conducts interactive computing through Jupyter Notebook.

## 2. R

It won't come as a surprise to many that those who love statistics love R. The "language of statistics" as it is popularly called, R is used to build data models which can be used for effective and accurate data analysis.

Powered by a large repository of R packages (CRAN, also called Comprehensive R Archive Network), with R you have just about every type of tool to accomplish any task in big data processing—right from analysis to data visualization. R can be integrated seamlessly with Apache Hadoop and Apache Spark, among other popular frameworks, for big data processing and analytics.

One issue with using R as a programming language for big data is that it is not very general-purpose. It means the code written in R is not production-deployable and generally has to be translated to some other programming language such as Python or Java. That said, if your goal is to only build statistical models for big data analytics, R is an option you should definitely consider.

Why do we use R for big data?

- It is built for data science.
- It supports Hadoop and Spark.
- It has strong statistical modeling and visualization capabilities.
- It supports Jupyter Notebook.

## 3. Java

Some of the traditional big data frameworks such as Apache Hadoop and all the tools within its ecosystem are all Java-based, and they are still in use today in many enterprises. Not to mention the fact that Java is one of the most stable and production-ready language among all the languages.

Using Java to develop your big data applications gives you the ability to use a large ecosystem of tools and libraries for interoperability, monitoring and much more, most of which have already been tried and tested.

One major drawback of Java is its verbosity. The fact that you have to write hundreds of lines of codes in Java for a task which can written in barely 15-20 lines of code in Python or Scala, can turnoff many budding programmers. However, the introduction of lambda functions in Java 8 does make life quite easier. Unlike newer languages like Python, Java does not support iterative development, and this is an area of focus for the future Java releases.

Despite the flaws, Java remains a strong contender when it comes to the preferred language for big data programming.

Why do we use Java for big data?

- Traditional big data tools and frameworks are written in Java.
- It is stable and production-ready.
- It is a large ecosystem of tried and tested tools and libraries.

## 4. Go

Go is one of the fastest rising programming languages in recent times. Designed by a group of Google engineers who were not very comfortable with C++, we think Go is a good shout in this list, simply because of the fact that it powers so many tools used in the big data infrastructure, including Kubernetes, Docker and many more.

Go is fast, easy to learn and to use. More importantly, as businesses look at building data analysis systems that can operate at scale, Go-based systems are being used to integrate machine learning and parallel processing of data. It is also possible to interface other languages with Go-based systems with relative ease.

Why do we use Go for big data?

● It is fast and easy to use.

● Many tools used in the big data infrastructure are Go-based.

● It has efficient distributed computing.

## 5. Scala

Last but not least is Scala. A beautiful crossover of the object-oriented and functional programming paradigms, Scala is fast and robust, and it is a popular choice of language for many big data professionals.The fact that two of the most popular big data processing frameworks in Apache Spark and Apache Kafka have been built on top of Scala tells you everything you need to know about the power of Scala.

Scala runs on the JVM, which means the codes written in Scala can be easily used within a Java-based big data ecosystem. One significant factor that differentiates Scala from Java, though, is that Scala is a lot less verbose in comparison. You can write hundreds of lines of confusing-looking Java code in less than 15 lines in Scala. One negative aspect of Scala, though, is its steep learning curve when compared to languages like Go and Python, and this may put off beginners looking to use it.

Why do we use Scala for big data?

● It is fast and robust.

● It is suitable for working with big data tools like Apache Spark for distributed big data processing.

● It is JVM compliant and can be used in a Java-based ecosystem.

There are a few other languages you might want to consider—Julia, SAS and MATLAB being some major ones which are useful in their own right. However, when compared to the languages we talked about above, we thought they fell a bit short in some aspects—be it speed, efficiency, ease of use, documentation or community support, among other things.

Now comes the question: which language should you choose?

It all depends on the which you want to do. If your focus is hardcore data analysis which involves a lot of statistical computing, R would be your go-to language. If you want to develop streaming applications for your big data, Scala can be a preferable choice. If you wish to use machine learning and build predictive models, Python will come to your rescue. Lastly, if you plan to build big data solutions using just the traditionally-available tools, Java is the language for you.

You also have the option of combining the power of two languages to get a more efficient and powerful solution. For example, you can train your machine learning model in Python and deploy it on Spark in a distributed mode. Ultimately, it all depends on how efficiently your solution can function, and more importantly is how fast and accurate it is.

## ✎ New Words

| | | |
|---|---|---|
| scene | [si:n] | n.地点，现场；场面 |
| framing | ['freɪmɪŋ] | n.构架；框架 |
| architecture | ['ɑ:kɪtektʃə] | n.体系结构；（总体、层次）结构 |
| spectrum | ['spektrəm] | n.光谱，波谱；范围，系列 |
| library | ['laɪbrəri] | n.库 |
| ecosystem | ['i:kəʊsɪstəm] | n.生态系统 |
| interactive | [ˌɪntər'æktɪv] | adj.交互式的；互动的 |
| effortlessly | ['efətləsli] | adv.轻松地，不费力地 |
| comprehensive | [ˌkɒmprɪ'hensɪv] | adj.广泛的；综合的 |
| accomplish | [ə'kʌmplɪʃ] | v.完成，达成 |
| seamlessly | ['si:mləsli] | adv.无缝地，无空隙地 |
| deployable | [dɪ'plɔɪ'eɪbl] | adj.可部署的 |
| definitely | ['defɪnətli] | adv.确定；明显地；明确地 |
| interoperability | [ˌɪntərˌɒpərə'bɪləti] | n.互用性，协同工作的能力 |
| verbosity | [vɜ:'bɒsəti] | n.冗长，赘言 |
| turnoff | ['tɜ:nɒf] | n.避开；岔开 |
| flaw | [flɔ:] | n.瑕疵，缺点 |
| contender | [kən'tendə] | n.（冠军）争夺者，竞争者 |
| frustrated | [frʌ'streɪtɪd] | adj.挫败的，失意的，泄气的 |
| crossover | ['krɒsəʊvə] | n.交叉 |
| significant | [sɪg'nɪfɪkənt] | adj.重要的；显著的 |
| verbose | [vɜ:'bəʊs] | adj.冗长的，啰嗦的 |
| hardcore | ['hɑ:dkɔ:] | n.核心部分 |
| rescue | ['reskju:] | v.营救，救助<br>n.营救（行动） |

## ✎ Phrases

| | |
|---|---|
| be declared as | 宣布为 |
| deep learning | 深度学习 |

| | |
|---|---|
| make up for | 弥补 |
| scripting language | 脚本语言 |
| iterative development | 迭代开发 |
| budding programmer | 新手程序员 |
| parallel processing | 并行处理 |
| distributed computing | 分布式计算 |
| learning curve | 学习曲线 |
| statistical computing | 统计计算 |
| streaming application | 流媒体应用 |
| distributed mode | 分布模式 |

## ✍ Abbreviations

| | |
|---|---|
| CRAN (Comprehensive R Archive Network) | 综合 R 档案网络 |
| JVM (Java Virtual Machine) | Java 虚拟机 |

## Reference Translation

## 用于大数据的五种编程语言

大数据专业人员必须做出的最重要的决定之一，尤其是对于那些刚进圈子或刚刚起步的人来说，就是为大数据操作和分析选择最佳的编程语言。如今，仅了解大数据问题并构筑架构以解决该问题还远远不够，也需要完美执行，而且选择正确的语言任重而道远。以下是大数据最常用的五种编程语言。

### 1. Python

根据最近举行的 Stack Overflow 开发人员调查，Python 已被宣布为增长最快的编程语言之一。它的通用性意味着它可以广泛使用于各种用例中，大数据编程是其应用的一个主要领域。

大数据框架中越来越多地使用许多用于数据分析和处理的库来清洗和处理大块数据，如 Pandas、NumPy、SciPy 都是基于 Python 的。不仅如此，大多数流行的机器学习和深度学习框架（如 Scikit-Learn、TensorFlow 等）也都使用 Python 编写，并且正在大数据生态系统中得到越来越多的应用。

使用 Python 的一个缺点是它运行缓慢，这也是为什么它在大数据编程方面还不是一流选择的原因。尽管非常易于使用，但大数据专业人员发现使用 Java 或 Scala 等语言构建的系统比使用 Python 构建的系统更快速、更强大。

但是，Python 用其他特性弥补了这一限制。由于 Python 主要是一种脚本语言，因此交互式编码和大数据分析解决方案的开发变得非常容易。Python 可以轻松地与现有的大数据框架（如 Apache Hadoop 和 Apache Spark）集成，从而使你能够大规模执行预测分析。

为什么我们将 Python 用于大数据？

● 它是通用的。

- 拥有丰富的数据分析和机器学习库。
- 它容易使用。
- 它支持迭代开发。
- 它集成了丰富的大数据工具。
- 通过 Jupyter Notebook 进行交互式计算。

## 2. R

喜欢统计的人一般都喜欢 R，很多人对此并不会感到惊讶。R 被普遍称为"统计语言"，用于建立有效而准确的数据分析的数据模型。

在大型 R 包存储库（CRAN，也称为综合 R 存档网络）的支持下，使用 R，你几乎拥有完成大数据处理中任何任务的所有类型的工具——从分析到数据可视化。R 可以与 Apache Hadoop 和 Apache Spark 以及其他流行框架无缝集成，用于大数据处理和分析。

使用 R 作为大数据编程语言的一个问题是它不是很通用。这意味着用 R 编写的代码不可用于产品部署，并且通常必须转换为某些其他编程语言，如 Python 或 Java。也就是说，如果你的目标只是为大数据分析构建统计模型，那么你绝对应该考虑使用 R。

为什么我们将 R 用于大数据？
- 它是为数据科学而构建的。
- 它支持 Hadoop 和 Spark。
- 它具有强大的统计建模和可视化功能。
- 它支持 Jupyter Notebook。

## 3. Java

一些传统的大数据框架（如 Apache Hadoop）及其生态系统中的所有工具都是基于 Java 的，如今它们仍在许多企业中使用。更不用说 Java 是所有语言中最稳定和最现成的语言之一。

使用 Java 开发大数据应用程序使你能够使用大型的工具和库生态系统来实现互操作性、监控以及更多功能，其中大多数已经过尝试和测试。

Java 的主要缺点之一是冗长。你必须用 Java 编写数百行代码来完成一项任务，而该任务用 Python 或 Scala 几乎只编写 15～20 行代码就可以了，这一事实可能会使许多新手程序员望而却步。但是，在 Java 8 中引入 lambda 函数确实使生活变得更加轻松。与 Python 等较新的语言不同，Java 不支持迭代开发，这是将来的 Java 版本关注的领域。

尽管 Java 存在缺陷，但在大数据编程的首选语言方面，它仍然是强大的竞争者。

为什么我们将 Java 用于大数据？
- 传统的大数据工具和框架是用 Java 编写的。
- 稳定且现成。
- 它是一个由久经考验的工具和库组成的大型生态系统。

## 4. Go

Go 是最近发展最快的编程语言之一。其由一群对 C++不太满意的谷歌工程师设计，我们认为 Go 在此列表中是一个不错的选择，这仅仅是因为它支持大数据基础架构中使用的许多工具，包括 Kubernetes、Docker 等。

Go 快速、易学且易用。更重要的是，随着企业着眼于构建可大规模运行的数据分析系统，基于 Go 的系统已用于集成机器学习和数据的并行处理。还可以相对轻松地将其他语言与基于 Go 的系统进行交互。

为什么我们将 Go 用于大数据？

- 快速且易于使用。
- 大数据基础架构中使用的许多工具都是基于 Go 的。
- 具有高效的分布式计算。

### 5. Scala

最后但并非最不重要的是 Scala。Scala 是面向对象和函数式编程范式的完美结合，它既快速又健壮，是许多大数据专业人士常用的语言选择。事实上，在 Apache Spark 和 Apache Kafka 中两个最受欢迎的大数据处理框架构建在 Scala 之上，这可以告诉你有关 Scala 功能的所有信息。

Scala 在 JVM 上运行，这意味着用 Scala 编写的代码可以在基于 Java 的大数据生态系统中轻松使用。不过，使 Scala 与 Java 不同的一个重要因素是，相比之下，Scala 要简洁得多。你可以在 Scala 中用不到 15 行来编写看上去令人困惑的数百行 Java 代码。但是，与 Go 和 Python 等语言相比，Scala 的不利方面是其陡峭的学习曲线，这可能会使初学者不愿使用它。

为什么我们将 Scala 用于大数据？

- 它快速而强大。
- 它适合与 Apache Spark 等大数据工具一起用于分布式大数据处理。
- 它适用于 JVM，可以在基于 Java 的生态系统中使用。

你可能还需要考虑其他几种语言——Julia、SAS 和 MATLAB 是一些主要的语言，它们本身就很有用。但是，与我们上面讨论的语言相比，我们认为它们在某些方面有所欠缺——无论在速度、效率、易用性、文档还是社区支持等方面。

现在出现的问题是：你应该选择哪种语言？

这完全取决于你要做什么。如果你的重点是涉及大量统计计算的核心数据分析，那么 R 将是你的首选语言。如果你想为大数据开发流应用程序，Scala 可能是一个更好的选择。如果你希望使用机器学习并构建预测模型，那么 Python 将助你一臂之力。最后，如果你打算仅使用传统上可用的工具来构建大数据解决方案，那么 Java 是适合你的语言。

你还可以选择结合两种语言的功能以获得更有效和强大的解决方案。例如，你可以使用 Python 训练你的机器学习模型，然后以分布式模式将其部署在 Spark 上。最终，这一切都取决于解决方案的运行效率，更重要的是它的速度和准确性如何。

# Text B

## Apache Spark

扫码听课文

Apache Spark is an open source data-processing engine for machine learning and AI applications, backed by the largest open source community in big data.

## 1. What is Apache Spark

Apache Spark is an open source data-processing engine for large data sets. It is designed to deliver the computational speed, scalability and programmability required for big data, specifically for streaming data, graph data, machine learning and Artificial Intelligence (AI) applications.

Spark's analytics engine processes data 10 to 100 times faster than alternatives. It scales by distributing processing work across large clusters of computers, with built-in parallelism and fault tolerance. It even includes APIs for programming languages that are popular among data analysts and data scientists, including Scala, Java, Python and R.

Spark is often compared with Apache Hadoop, and specifically with MapReduce, Hadoop's native data-processing component. The chief difference between Spark and MapReduce is that Spark processes data and keeps the data in memory for subsequent steps without writing to or reading from disk, which results in dramatically faster processing speeds.

## 2. How does Apache Spark work

Apache Spark has a hierarchical master/slave architecture. The Spark Driver is the master node that controls the cluster manager, which manages the worker (slave) nodes and delivers data results to the application client.

Based on the application code, Spark Driver generates the SparkContext, which works with the cluster manager (Spark's Standalone Cluster Manager or other cluster managers like Hadoop YARN, Kubernetes or Mesos) to distribute and monitor execution across the nodes. It also creates Resilient Distributed Datasets (RDD), which are the key to Spark's remarkable processing speed.

(1) Resilient Distributed Dataset (RDD)

Resilient distributed datasets are fault-tolerant collections of elements that can be distributed among multiple nodes in a cluster and worked on in parallel. RDD is a fundamental structure in Apache Spark.

Spark loads data by referencing a data source or by parallelizing an existing collection with the SparkContext parallelize method into a RDD for processing. Once data is loaded into a RDD, Spark performs transformations and actions on RDDs in memory, which is the key to Spark's speed. Spark also stores the data in memory unless the system runs out of memory or the user decides to write the data to disk for persistence.

Each dataset in a RDD is divided into logical partitions, which may be computed on different nodes of the cluster. Users can perform two types of RDD operations: transformations and actions. Transformations are operations applied to create a new RDD. Actions are used to instruct Apache Spark to apply computation and pass the result back to the driver.

Spark supports a variety of actions and transformations on RDDs. This distribution is done by Spark, so users don't have to worry about computing the right distribution.

(2) Directed Acyclic Graph (DAG)

As opposed to the two-stage execution process in MapReduce, Spark creates a Directed Acyclic Graph (DAG) to schedule tasks and the orchestration of worker nodes across the cluster. As Spark acts and transforms data in the task execution processes, the DAG scheduler facilitates efficiency by orchestrating the worker nodes across the cluster. This task-tracking makes fault tolerance possible, as it reapplies the recorded operations to the data from a previous state.

(3) DataFrames and Datasets

In addition to RDDs, Spark handles two other data types: DataFrames and Datasets.

DataFrames are the most common structured Application Programming Interfaces (API) and they represent a table of data with rows and columns. Although RDD has been a critical feature to Spark, it is now in maintenance mode. Because of the popularity of Spark's Machine Learning library (MLlib), DataFrames have taken on the lead role as the primary API for MLlib. This is important to note when using the MLlib API, as DataFrames provide uniformity across the different languages, such as Scala, Java, Python and R.

Datasets are an extension of DataFrames that provide a type-safe, object-oriented programming interface. Datasets are, by default, a collection of strongly typed JVM objects, unlike DataFrames.

Spark SQL allows data to be queried from DataFrames and SQL data stores, such as Apache Hive. Spark SQL queries return a DataFrame or Dataset when they are run within another language.

(4) Spark Core

Spark Core is the base for all parallel data processing and it handles scheduling, optimization, RDD and data abstraction. Spark Core provides the functional foundation for the Spark libraries, Spark SQL, Spark Streaming, the MLlib (machine learning library) and GraphX graph data processing. The Spark Core and cluster manager distribute data across the Spark cluster and abstract it. This distribution and abstraction make handling big data very fast and user-friendly.

(5) Spark APIs

Spark includes a variety of Application Programming Interfaces (API) to bring the power of Spark to the broadest audience. Spark SQL allows for interaction with RDD data in a relational manner. Spark also has a well-documented API for Scala, Java, Python and R. Each language's API in Spark has its specific nuances in how it handles data. RDDs, DataFrames and Datasets are all available in each language's API. With APIs in such a variety of languages, Spark makes big data processing accessible to more diverse groups of people with backgrounds in development, data science and statistics.

## 3. Apache Spark and machine learning

Spark has various libraries that extend the capabilities to machine learning, artificial intelligence (AI) and stream processing.

(1) Apache Spark MLlib

One of the critical capabilities of Apache Spark is the machine learning abilities available in the Spark MLlib. The Apache Spark MLlib provides an out-of-the-box solution for doing classification and regression, collaborative filtering, clustering, distributed linear algebra, decision trees, random forests, gradient-boosted trees, frequent pattern mining, evaluation metrics and statistics. The capabilities of the MLlib, combined with the various data types Spark can handle, make Apache Spark an indispensable big data tool.

(2) Spark GraphX

In addition to having API capabilities, Spark has Spark GraphX, a new addition to Spark designed to solve graph problems. GraphX is a graph abstraction that extends RDDs for graphs and graph-parallel computation. Spark GraphX integrates with graph databases that store interconnectivity information or webs of connection information, like that of a social network.

(3) Spark Streaming

Spark Streaming is an extension of the core Spark API that enables scalable and fault-tolerant processing of live data streams. As Spark Streaming processes data, it can deliver data to file systems, databases and live dashboards for real-time streaming analytics with Spark's machine learning and graph-processing algorithms. Built on the Spark SQL engine, Spark Streaming also allows for incremental batch processing that results in faster processing of streamed data.

**4. Spark vs. Apache Hadoop and MapReduce**

"Spark vs. Hadoop" is a frequently searched term on the web, but as noted above, Spark is more of an enhancement to Hadoop — and, more specifically, to Hadoop's native data processing component, MapReduce. In fact, Spark is built on the MapReduce framework, and today, most Hadoop distributions include Spark.

Like Spark, MapReduce enables programmers to write applications that process huge data sets faster by processing portions of the data set in parallel across large clusters of computers. While MapReduce processes data on disk, adding read and write times that slow processing, Spark performs calculations in memory, which is much faster. As a result, Spark can process data up to 100 times faster than MapReduce.

Spark's built-in APIs for multiple languages make it more practical and approachable for developers than MapReduce, which has a reputation for being difficult to program. Unlike MapReduce, Spark can run stream-processing applications on Hadoop clusters using YARN, Hadoop's resource management and job scheduling framework. As noted above, Spark adds the capabilities of MLlib, GraphX and SparkSQL. And Spark can handle data from other data sources outside of the Hadoop application, including Apache Kafka.

In addition, Spark is compatible with and complementary to Hadoop. It can process Hadoop data, including data from HDFS (Hadoop Distributed File System), HBase (a non-relational database that runs on HDFS), Apache Cassandra (a NoSQL alternative to HDFS) and Hive (a Hadoop-based data warehouse).

## ✍ New Words

| | | |
|---|---|---|
| community | [kə'mju:nəti] | n.社团，社区 |
| computational | [ˌkɒmpju'teɪʃənl] | adj.计算的 |
| programmability | [ˌprəʊɡræmə'bɪlɪti] | n.可编程性 |
| parallelism | ['pærəlelɪzəm] | n.平行；对应；类似 |
| specifically | [spə'sɪfɪkli] | adv.特有地，明确地 |

| subsequent | ['sʌbsɪkwənt] | adj.后来的，随后的 |
| dramatically | [drə'mætɪk(ə)li] | adv.戏剧性地，引人注目地；显著地 |
| remarkable | [rɪ'mɑːkəbl] | adj.卓越的；显著的 |
| persistence | [pə'sɪstəns] | n.持久性，持久化 |
| reapply | [ˌriːə'plaɪ] | v.再运用；再申请 |
| state | [steɪt] | n.状态 |
| uniformity | [ˌjuːnɪ'fɔːməti] | n.同一性；同样，一样 |
| optimization | [ˌɒptɪmaɪ'zeɪʃn] | n.最佳化，最优化 |
| user-friendly | [juːzə'frendli] | adj.用户友好的，用户容易掌握使用的 |
| nuance | ['njuːɑːns] | n.细微差别 |
| collaborative | [kə'læbərətɪv] | adj.合作的，协作的 |
| random | ['rændəm] | adj.任意的；随机的 |
| metrics | ['metrɪks] | n.指标；度量 |
| indispensable | [ˌɪndɪ'spensəbl] | adj.不可缺少的，绝对必要的 |
| interconnectivity | [ɪntəkənek'tɪvɪti] | n.互联性；互联互通；连通性 |
| enhancement | [ɪn'hɑːnsmənt] | n.增强；提高；改善 |
| approachable | [ə'prəʊtʃəbl] | adj.可亲近的；可接近的 |
| complementary | [ˌkɒmplɪ'mentri] | adj.互补的；补充的，补足的 |

## ✎ Phrases

| data-processing engine | 数据处理引擎 |
| open source community | 开源社区 |
| streaming data | 流数据 |
| graph data | 图形数据 |
| data analyst | 数据分析师 |
| data scientist | 数据科学家 |
| master/slave architecture | 主/从结构 |
| data source | 数据源 |
| be divided into | 被分为 |
| logical partition | 逻辑分区 |
| data type | 数据类型 |
| maintenance mode | 维护模式 |
| parallel data processing | 并行数据处理 |
| out-of-the-box solution | 开箱即用的解决方案 |
| collaborative filtering | 协同过滤 |
| linear algebra | 线性代数 |

| random forest | 随机森林 |
| gradient-boosted tree | 梯度引导树 |
| frequent pattern mining | 频繁模式挖掘 |
| evaluation metrics | 评价指标 |
| combine with | 与……结合 |
| graph-parallel computation | 图并行计算 |
| graph-processing algorithm | 图处理算法 |

## Abbreviations

| API (Application Programming Interface) | 应用程序编程接口 |
| RDD (Resilient Distributed Dataset ) | 弹性分布式数据集 |
| DAG (Directed Acyclic Graph) | 有向无环图 |
| MLlib (Machine Learning library) | 机器学习库 |

# Reference Translation

# Apache Spark 软件

Apache Spark 是用于机器学习和 AI 应用程序的开源数据处理引擎，并由最大的大数据开源社区提供支持。

## 1. 什么是 Apache Spark

Apache Spark 是用于大型数据集的开源数据处理引擎。它旨在提供大数据（特别是流数据、图形数据、机器学习和人工智能应用程序）所需的计算速度、可扩展性和可编程性。

Spark 的分析引擎处理数据的速度比其他的引擎快 10～100 倍。它具有内置的并行性和容错能力，可通过在大型计算机集群中分配处理工作来扩展规模。它甚至包括用于编程语言的 API，这些编程语言在数据分析人员和数据科学家中很流行，包括 Scala、Java、Python 和 R。

通常将 Spark 与 Apache Hadoop 进行比较，尤其是与 Hadoop 的本地数据处理组件 MapReduce 进行比较。Spark 和 MapReduce 之间的主要区别在于，Spark 处理数据并将数据保留在内存中以供后续步骤使用，而无须写入磁盘或从磁盘读取数据，从而大大加快了处理速度。

## 2. Apache Spark 如何工作

Apache Spark 具有分层的主/从体系结构。Spark Driver 是控制集群管理器的主节点，集群管理器管理工作器（从属）节点并将数据结果传递给应用程序客户端。

Spark Driver 根据应用程序代码生成 SparkContext，SparkContext 可与集群管理器（Spark 的独立集群管理器或 Hadoop YARN、Kubernetes 或 Mesos 等其他集群管理器）一起使用，以在节点之间分发和监控执行。它还创建了弹性分布式数据集（RDD），这是 Spark 处理速度优异的关键。

（1）弹性分布式数据集（RDD）

弹性分布式数据集是元素的容错集合，可以在集群中的多个节点之间分布且并行处理。RDD 是 Apache Spark 中的基础结构。

Spark 通过引用数据源或通过使用 SparkContext 并行化方法将现有集合并行到 RDD 中进行处理来加载数据。将数据加载到 RDD 中后，Spark 会对内存中的 RDD 执行转换和操作，这是 Spark 速度快的关键。Spark 还会将数据存储在内存中，除非系统内存不足或用户决定将数据写入磁盘以实现持久性存储。

RDD 中的每个数据集都被划分为逻辑分区，可以在集群的不同节点上进行计算。用户可以执行两种类型的 RDD 操作：转换和行动。转换是用于创建新 RDD 的操作。行动用于指示 Apache Spark 进行计算并将结果传递回驱动器。

Spark 支持 RDD 上的许多转换和行动。该分布是由 Spark 完成的，因此用户无须考虑正确地分布计算。

（2）有向无环图（DAG）

与 MapReduce 中的两阶段执行过程相反，Spark 创建一个有向无环图（DAG）来调度任务和跨集群的工作节点的编排。当 Spark 在任务执行过程中行动和转换数据时，DAG 调度程序通过在整个集群中协调工作节点来提高效率。这种任务跟踪使容错成为可能，因为它会将记录的操作重新应用于先前状态的数据。

（3）DataFrame 和 Dataset

除了 RDD 之外，Spark 还处理其他两种数据类型：DataFrame 和 Dataset。

DataFrame 是最常见的结构化应用程序编程接口（API），它们表示具有行和列的数据表。尽管 RDD 一直是 Spark 的关键功能，但现在处于维护模式。由于 Spark 的机器学习库（MLlib）的普及，DataFrame 作为 MLlib 的主要 API 发挥了主导作用。使用 MLlib API 时需要注意这一点，因为 DataFrame 提供了不同语言（如 Scala、Java、Python 和 R）的统一性。

Dataset 是 DataFrame 的扩展，提供了类型安全的、面向对象的编程接口。默认情况下，Dataset 是强类型 JVM 对象的集合，这与 DataFrame 不同。

Spark SQL 允许从 DataFrame 和 SQL 数据存储（如 Apache Hive）中查询数据。当以其他语言运行时，Spark SQL 查询将返回 DataFrame 或 Dataset。

（4）Spark Core

Spark Core 是所有并行数据处理的基础，并处理调度、优化、RDD 和数据抽象。Spark Core 为 Spark 库、Spark SQL、Spark Streaming、MLlib（机器学习库）和 GraphX 图形数据处理提供了功能基础。Spark Core 和集群管理器将数据分布在整个 Spark 集群中并对其进行抽象。这种分布和抽象使处理大数据变得非常快速而且对用户友好。

（5）Spark API

Spark 包含各种应用程序编程接口（API），可将 Spark 的功能带给最广泛的受众。Spark SQL 允许以关系方式与 RDD 数据进行交互。Spark 还具有针对 Scala、Java、Python 和 R 的文档丰富的 API。Spark 中每种语言的 API 在处理数据方面都有其特定的细微差别。RDD、DataFrame 和 Datasets 在每种语言的 API 中都可以使用。通过使用多种语言的 API，Spark 使得具有开发、数据科学和统计背景的不同人群可以访问大数据处理。

### 3. Apache Spark 和机器学习

Spark 拥有各种库，这些库将功能扩展到机器学习、人工智能（AI）和流媒体处理。

(1) Apache Spark MLlib

Apache Spark 的关键功能之一是 Spark MLlib 中提供的机器学习功能。Apache Spark MLlib 提供了一种开箱即用的解决方案，用于进行分类和回归、协作过滤、聚类、分布式线性代数、决策树、随机森林、梯度增强树、频繁模式挖掘、评估指标和统计信息。MLlib 的功能与 Spark 可以处理的各种数据类型相结合，使 Apache Spark 成为必不可少的大数据工具。

(2) Spark GraphX

除了具有 API 功能外，Spark 还具有 Spark GraphX，这是 Spark 的新增的功能，旨在解决图形问题。GraphX 是一种图形抽象，它扩展了 RDD 用于图形和图形并行计算的功能。Spark GraphX 与图数据库集成，该图数据库存储互联信息或连接信息的网络，如社交网络的信息。

(3) Spark Streaming

Spark Streaming 是核心 Spark API 的扩展，可实现实时数据流的可扩展、容错处理。在 Spark Streaming 处理数据时，它可以用 Spark 的机器学习和图形处理算法将数据传递到文件系统、数据库和实时仪表板，以进行实时流分析。Spark Streaming 基于 Spark SQL 引擎构建，还允许增量批处理，从而可以更快地处理流数据。

### 4. Spark 与 Apache Hadoop 和 MapReduce

"Spark vs Hadoop" 是网络上经常搜索的词语，但如上所述，Spark 是 Hadoop 的增强版——更具体地说，是对 Hadoop 的本机数据处理组件 MapReduce 的增强。实际上，Spark 是基于 MapReduce 框架构建的，如今大多数 Hadoop 发行版都包含 Spark。

与 Spark 一样，MapReduce 使程序员能够编写应用程序，这些程序能够通过在大型计算机集群并行处理部分数据集来更快地处理大量数据集。MapReduce 处理磁盘上的数据，这就增加了读取和写入次数，减慢了处理速度，而 Spark 在内存中执行计算要快得多。因此，Spark 处理数据的速度比 MapReduce 快 100 倍。

与 MapReduce 相比，Spark 内置的针对多种语言的 API 使它对开发人员更实用、更易上手，因为 MapReduce 以难以编程而著称。与 MapReduce 不同，Spark 可以使用 YARN（Hadoop 的资源管理和作业调度框架）在 Hadoop 集群上运行流处理应用程序。如上所述，Spark 添加了 MLlib、GraphX 和 SparkSQL 的功能。而且 Spark 可以处理 Hadoop 应用程序之外的其他数据源（包括 Apache Kafka）中的数据。

此外，Spark 与 Hadoop 兼容并互补。它可以处理 Hadoop 数据，包括来自 HDFS（Hadoop 分布式文件系统）、HBase（在 HDFS 上运行的非关系数据库）、Apache Cassandra（HDFS 的 NoSQL 替代品）和 Hive（基于 Hadoop 的数据仓库）中的数据。

# Exercises

[Ex. 1] 根据 Text A 回答以下问题。

1. What is one of the most important decisions that big data professionals have to make, especially the ones who are new to the scene or are just starting out?

2. What has Python been declared as? What does its general-purpose nature mean?

3. What can Python integrate effortlessly with?

4. What is R popularly called? What is R used to do?

5. What is one issue with using R as a programming language for big data? What does it mean?

6. What abilities does using Java to develop your big data applications give you?

7. Why do we use Java for big data?

8. What is Go? Why do you think Go is a good shout in this list?

9. Why do we use Go for big data?

10. What is one significant factor that differentiates Scala from Java?

[Ex. 2] 根据 Text B 回答以下问题。

1. What is Apache Spark? What is it designed to do?

2. What is the chief difference between Spark and MapReduce?

3. What is the Spark Driver?

4. What are Resilient Distributed Datasets (RDDs)?

5. How does the DAG scheduler facilitate efficiency?

6. Why have DataFrames taken on the lead role as the primary API for MLlib?

7. What is Spark Core? What does it handle?

8. What is one of the critical capabilities of Apache Spark? What does the Apache Spark MLlib provide?

9. What is GraphX? What does Spark GraphX integrate with?

10. What does MapReduce enable programmers to do? What is the difference between MapReduce and Spark?

**[Ex. 3]** 词汇英译中

| | |
|---|---|
| 1. deep learning | 1. _____ |
| 2. accomplish | 2. _____ |
| 3. distributed computing | 3. _____ |
| 4. ecosystem | 4. _____ |
| 5. learning curve | 5. _____ |
| 6. architecture | 6. _____ |
| 7. distributed mode | 7. _____ |
| 8. interactive | 8. _____ |
| 9. parallel processing | 9. _____ |
| 10. seamlessly | 10. _____ |

**[Ex. 4]** 词汇中译英

| | |
|---|---|
| 1. 脚本语言 | 1. _____ |
| 2. 互补的；补充的，补足的 | 2. _____ |
| 3. 流媒体应用 | 3. _____ |
| 4. 增强；提高；改善 | 4. _____ |
| 5. 数据源 | 5. _____ |
| 6. 最佳化，最优化 | 6. _____ |
| 7. 数据类型 | 7. _____ |
| 8. 平行；对应；类似 | 8. _____ |
| 9. 随机森林 | 9. _____ |
| 10. 同一性；同样，一样 | 10. _____ |

**[Ex. 5]** 短文翻译

**Python for Data Science**

Python is an interpreted, interactive and object-oriented programming. It is one of the best languages used by data scientist for various data sciences projects/application. Python provides great functionality to deal with mathematics, statistics and scientific function. It provides great libraries to deals with data science application.

One of the main reasons why Python is widely used in the scientific and research communities is because of its ease of use and simple syntax, which makes it easy to adapt for people who do not have an engineering background. It is also more suited for quick prototyping.

According to engineers coming from academia and industry, deep learning frameworks available with Python APIs, in addition to the scientific packages have made Python incredibly productive and versatile. There has been a lot of evolution in deep learning Python frameworks and it's rapidly upgrading.

In terms of application areas, ML scientists prefer Python as well. When it comes to areas like building fraud detection algorithms and network security, developers leaned towards Java, while for applications like natural language processing (NLP) and sentiment analysis, developers opted for Python, because it provides large collection of libraries that help to solve complex business problem easily, build strong systems and data application.

The following are some useful features of Python language.

- It uses the elegant syntax, hence the programs are easier to read.
- It is a simple to access language, which makes it easy to achieve the program working.
- It has large standard libraries and community support.
- The interactive mode of Python makes it simple to test codes.
- In Python, it is also simple to extend the code by appending new modules that are implemented in other compiled language like C++ or C.
- Python is an expressive language which is possible to embed into applications to offer a programmable interface.
- It allows developers to run the code anywhere, including Windows, Mac OS X, UNIX and Linux.
- It is free software. It does not cost anything to use or download Python or to add it to the application.

# Reading Material

## Hadoop

Today tons of companies are adopting Hadoop big data tools to solve their big data queries and their customer market segments.

### 1. Hadoop consists of mainly 3 components

- HDFS (Hadoop Distributed File System): HDFS is working as a storage layer on Hadoop. The data is always stored in the form of data blocks[1] on HDFS where the default size of each data block is 128MB in size which is configurable[2]. Hadoop works on the MapReduce algorithm which is a master-slave architecture[3]. HDFS has NameNode and DataNode that work in a similar pattern.
- MapReduce: MapReduce works as a processing layer on Hadoop. MapReduce is a programming model that is mainly divided into two phases MapPhase and Reduce Phase. It is designed for processing the data in parallel[4] which is divided on various machines (nodes).

---

1  data block: 数据块
2  configurable [kən'fiɡərəbl] adj.可配置的
3  master-slave architecture: 主从结构
4  parallel ['pærəlel] adj.平行的

● YARN (Yet Another Resources Negotiator): YARN is the job scheduling[5] and resource management layer in Hadoop. The data stored on HDFS is processed and run with the help of data processing engines like graph processing, interactive processing, batch processing[6], etc. The overall performance of Hadoop is improved up with the help of this YARN framework.

## 2. Features of Hadoop

(1) Open source

Hadoop is open source, which means it is free to use. Since it is an open source project the source code[7] is available online for anyone to understand it or make some modifications[8] as per their industry requirement.

(2) Highly scalable cluster

Hadoop is a highly scalable model. A large amount of data is divided into multiple inexpensive machines in a cluster which is processed parallelly. The number of these machines or nodes can be increased or decreased as per the enterprise's requirements. In traditional RDBMS (Relational DataBase Management System) the systems can not be scaled to approach large amounts of data.

(3) Fault tolerance[9] is available

Hadoop uses commodity hardware (inexpensive systems) which can be crashed at any moment. In Hadoop data is replicated[10] on various DataNodes in a Hadoop cluster which ensures the availability of data if somehow any of your systems got crashed. You can read all of the data from a single machine. If this machine faces a technical issue, data can also be read from other nodes in a Hadoop cluster because the data is copied or replicated by default. By default, Hadoop makes 3 copies of each file block and store it into different nodes. This replication factor is configurable and can be changed by changing the replication property in the hdfs-site.xml file.

(4) High availability is provided

Fault tolerance provides high availability in the Hadoop cluster. High availability means the availability of data on the Hadoop cluster. Due to fault tolerance in case if any of the DataNode goes down the same data can be retrieved from any other node where the data is replicated. The high available Hadoop cluster also has 2 or more than two Name Node, i.e. Active[11] NameNode and Passive[12] NameNode. In case the Active NameNode fails then the Passive NameNode will take the responsibility of Active NameNode and provide the same data as that of Active NameNode which can easily be utilized by the user.

(5) Cost effective

Hadoop is open source and it uses cost effective commodity hardware which provides a cost-

---

5　job scheduling: 作业调度

6　batch processing: 批处理

7　source code: 源代码

8　modification [ˌmɒdɪfɪˈkeɪʃn] *n.*修改

9　fault tolerance: 容错

10　replicate [ˈreplɪkeɪt] *vt.*复制

11　active [ˈæktɪv] *adj.*主动的

12　passive [ˈpæsɪv] *adj.*被动的

efficient model, unlike traditional relational databases that require expensive hardware and high end processors[13] to deal with big data. The problem with traditional relational databases is that storing the massive volume of data is not cost effective, so the company starts to remove the raw data[14], which may not result in the correct scenario of their business. Means Hadoop provides 2 main benefits with the cost one is it's open-source means free to use and the other is that it uses commodity hardware which is also inexpensive.

(6) Hadoop provide flexibility

Hadoop is designed in such a way that it can deal with any kind of dataset like structured (MySQL data), semistructured (XML, JSON) and unstructured (images and videos) very efficiently. This means it can easily process any kind of data independent of its structure, which makes it highly flexible. It is very useful for enterprises as they can process large datasets easily, so the businesses can use Hadoop to analyze valuable insights of data from sources like social media, email, etc. With this flexibility, Hadoop can be used with log processing, data warehousing, fraud detection, etc.

(7) Easy to use

Hadoop is easy to use since the developers don't need to worry about any of the processing work since it is managed by the Hadoop itself. Hadoop ecosystem[15] is also very large which comes up with lots of tools like Hive, Pig, Spark, HBase, Mahout, etc.

(8) Hadoop uses data locality[16]

The concept of data locality is used to make Hadoop processing fast. In the data locality concept, the computation logic is moved near data rather than moving the data to the computation logic. The cost of moving data on HDFS is costliest and with the help of the data locality concept, the bandwidth[17] utilization[18] in the system is minimized.

(9) Provides faster data processing

Hadoop uses a distributed file system to manage its storage, i.e. HDFS. In DFS (Distributed File System) a large size file is broken into small size file blocks then distributed among the Nodes available in a Hadoop cluster. This massive number of file blocks are processed parallelly, which makes Hadoop faster. HDFS provides a high-level performance as compared to the traditional DataBase Management Systems.

---

13    processor ['prəʊsesə] *n.*处理器

14    raw data: 原始数据

15    ecosystem ['i:kəʊsɪstəm] *n.*生态系统

16    locality [ləʊ'kæləti] *n.*局部性

17    bandwidth ['bændwɪdθ] *n.*带宽

18    utilization [ˌju:təlaɪ'zeɪʃn] *n.*利用，效用

# Unit 8

## Text A

### Data Visualization

扫码听课文

If you're very familiar with data analysis, then you've encountered data visualization. It is a key part of data analysis.

**1. What is data visualization and why is it important**

Data visualization is the representation of data or information in a graph, chart or other visual format. It communicates relationships of the data with images. It is important because it allows trends and patterns to be more easily seen. With the rise of big data, we need to be able to interpret increasingly larger batches of data. Machine learning makes it easier to conduct analyses such as predictive analysis. Data visualization is not only important for data scientists and data analysts, it is necessary to understand data visualization in any career. Whether you work in finance, marketing, tech, design or anything else, you need to visualize data.

**2. Why do we need data visualization**

We need data visualization because a visual summary of information makes it easier to identify patterns and trends than looking through thousands of rows on a spreadsheet. It's the way the human brain works. Since the purpose of data analysis is to gain insights, data is much more valuable when it is visualized. Even if a data analyst can pull insights from data without visualization, it will be more difficult to communicate the meaning without visualization. Charts and graphs make communicating data findings easier even if you can identify the patterns without them.

**3. What is data visualization used for**

(1) Changing over time

This is perhaps the most basic and common use of data visualization, but that doesn't mean it's not valuable. The reason is most data has an element of time involved. Therefore, the first step in a lot of data analyses is to see how the data trends over time.

(2) Determining frequency

Frequency is also a fairly basic use of data visualization because it also applies to data that involves time. If time is involved, it is logical that you should determine how often the relevant events happen over time.

(3) Determining relationships (correlations)

Identifying correlations is an extremely valuable use of data visualization. It is extremely difficult to determine the relationship between two variables without a visualization. It is important

*141*

to be aware of relationships in data. This is a great example of the value of data visualization in data analysis.

(4) Examining a network

An example of examining a network with data visualization can be seen in market research. Marketing professionals need to know which audiences to target with their message, so they analyze the entire market to identify audience clusters, bridges between the clusters, influencers within clusters and outliers.

(5) Scheduling

When planning out a schedule or timeline for a complex project, things can get confusing. A Gantt chart addresses that issue by clearly illustrating each task within the project and how long it will take to complete.

(6) Analyzing value and risk

Determining complex metrics such as value and risk requires many different variables to be factored in, making it almost impossible to see accurately with a plain spreadsheet. Data visualization can be as simple as color-coding a formula to show which opportunities are valuable and which are risky.

## 4. Types of data visualization charts

Now that we understand how data visualization can be used, let's apply the different types of data visualization to their uses. There are numerous tools available to help create data visualizations. Some are manual and some are automated, but either way they should allow you to make any of the following types of visualizations.

(1) Line chart

A line chart illustrates changes over time. The x-axis is usually a period of time, while the y-axis is quantity. So, this could illustrate a company's sales for the year broken down by month or how many units a factory produced each day for the past week.

(2) Area chart

An area chart is an adaptation of a line chart where the area under the line is filled in to emphasize its significance. The color fill for the area under each line should be somewhat transparent so that overlapping areas can be discerned.

(3) Bar chart

A bar chart illustrates changes over time. But if there is more than one variable, a bar chart can make it easier to compare the data for each variable at each moment in time. For example, a bar chart could compare the company's sales from this year to last year.

(4) Histogram

A histogram looks like a bar chart, but measures frequency rather than trends over time. The x-axis of a histogram lists the "bins" or intervals of the variable, and the y-axis is frequency, so each bar represents the frequency of that bin. For example, you could measure the frequencies of each answer to a survey question. The bins would be the answer "unsatisfactory" "neutral" and "satisfactory". This would tell you how many people there are for each answer.

(5) Scatter plot

Scatter plots are used to find correlations. Each point on a scatter plot means "when x = this, then y equals this". That way, if the points trend a certain way, there is a relationship between them. If the plot is truly scattered with no trend at all, then the variables do not affect each other at all.

(6) Bubble chart

A bubble chart is an adaptation of a scatter plot, where each point is illustrated as a bubble whose area has meaning in addition to its placement on the axes. A pain point associated with bubble charts is the limitations on sizes of bubbles due to the limited space within the axes. So, not all data will fit effectively in this type of visualization.

(7) Pie chart

A pie chart is the best option for illustrating percentages because it shows each element as part of a whole. So, if your data explains a breakdown in percentages, a pie chart will clearly present the pieces in the proper proportions.

(8) Gauge

A gauge can be used to illustrate the distance between intervals. This can be presented as a round clock-like gauge or as a tube type gauge resembling a liquid thermometer. Multiple gauges can be shown next to each other to illustrate the difference between multiple intervals.

(9) Map

Much of the data dealt with in businesses has a location element, which makes it easy to illustrate on a map. An example of a map visualization is mapping the number of purchases customers made in each state. In this example, each state would be shaded in and states with less purchases would be a lighter shade, while states with more purchases would be darker shades. Location information can also be very valuable for business leadership to understand, making this an important data visualization to use.

(10) Heat map

A heat map is basically a color-coded matrix. A formula is used to color each cell of the matrix to represent the relative value or risk of that cell. Usually colors of a heat map range from green to red, with green being a better result and red being worse. This type of visualization is helpful because colors are quicker to interpret than numbers.

(11) Frame diagram

Frame diagrams are basically tree maps which clearly show hierarchical relationship structure. A frame diagram consists of branches, which each have more branches connecting to them with each level of the diagram consisting of more and more branches.

## 5. Conclusion

Effective data visualization is the crucial final step of data analysis. Without it, important insights and messages can be lost.

# 🔊 New Words

| | | |
|---|---|---|
| visualization | [ˌvɪʒuəlaɪˈzeɪʃn] | n.可视化，形象化；虚拟化 |
| encounter | [ɪnˈkaʊntə] | vt.不期而遇；遭遇 |
| chart | [tʃɑːt] | n.图表<br>v.记录；绘制……的地图；制订（计划） |
| communicate | [kəˈmjuːnɪkeɪt] | vt.传达，表达；沟通；传递 |
| batch | [bætʃ] | n.一批<br>v.分批处理 |
| career | [kəˈrɪə] | n.生涯；经历<br>adj.就业的 |
| attention | [əˈtenʃn] | n.注意，注意力 |
| frequency | [ˈfriːkwənsi] | n.频繁性，频率 |
| relevant | [ˈreləvənt] | adj.有关的，相关联的 |
| influencer | [ˈɪnfluənsə] | n.意见领袖；有影响力的人 |
| timeline | [ˈtaɪmlaɪn] | n.时间轴，时间表 |
| confuse | [kənˈfjuːz] | vt.使混乱；使困惑；使难理解 |
| illustrate | [ˈɪləstreɪt] | vt.给……加插图；（用示例、图画等）说明 |
| formula | [ˈfɔːmjələ] | n.公式，准则 |
| opportunity | [ˌɒpəˈtjuːnəti] | n.机会；时机 |
| manual | [ˈmænjuəl] | adj.手动的，手工的<br>n.使用手册，说明书 |
| emphasize | [ˈemfəsaɪz] | v.（使）突出/明显；强调；重视 |
| significance | [sɪɡˈnɪfɪkəns] | n.重要性；含义 |
| discern | [dɪˈsɜːn] | v.辨别，分清 |
| histogram | [ˈhɪstəɡræm] | n.柱状图；直方图 |
| interval | [ˈɪntəvl] | n.（时间上的）间隔，区间 |
| bin | [bɪn] | n.区间；端点，范围 |
| survey | [ˈsɜːveɪ] | n.调查；概述<br>v.审察，测量；进行民意测验 |
| unsatisfactory | [ˌʌnˌsætɪsˈfæktəri] | adj.不能令人满意的；不符合要求的；不满足的 |
| adaptation | [ˌædæpˈteɪʃn] | n.改编版；适应 |
| percentage | [pəˈsentɪdʒ] | n.百分比 |
| shade | [ʃeɪd] | n.阴影；色度 |
| matrix | [ˈmeɪtrɪks] | n.矩阵；模型 |

## ✍ Phrases

| | |
|---|---|
| be familiar with | 熟悉的；友好的 |
| be aware of | 知道，意识到 |
| Gantt chart | 甘特图，线条图 |
| be factored in | 被考虑进去 |
| line chart | 线形图，折线图 |
| area chart | 面积图 |
| bar chart | 条形图 |
| scatter plot | 散点图，散布图 |
| bubble chart | 气泡图 |
| pain point | 痛点 |
| pie chart | 饼图 |
| heat map | 热图 |
| frame diagram | 框架图 |

## Reference Translation

## 数据可视化

如果你非常熟悉数据分析，那么会遇到数据可视化的问题。它是数据分析的关键部分。

### 1. 什么是数据可视化以及它为什么重要

数据可视化是以图形、图表或其他可视格式展示数据或信息。它用图像来表达数据之间的关系。它很重要，因为它使趋势和模式更容易看到。随着大数据的兴起，我们需要能够解释越来越大批的数据。机器学习使进行诸如预测分析之类的分析变得更容易。数据可视化不仅对数据科学家和数据分析人员很重要，而且在各行各业中都有助于数据的呈现。无论你从事金融、市场营销、技术、设计或其他领域的工作，都需要可视化数据。

### 2. 为什么需要数据可视化

我们需要数据可视化，因为与查看电子表格中数千行数据相比，信息的可视化摘要使识别模式和趋势更容易。这是人脑运作的方式。由于数据分析的目的是获取见解，因此在可视化数据时，其价值将大大提高。即使数据分析人员可以在没有可视化的情况下从数据中获取见解，但在没有可视化的情况下传达其含义将更加困难。尽管你不使用图表和图形也可以识别模式，但它们使交流数据发现变得更加容易。

### 3. 数据可视化的用途是什么

（1）随时间而变化

这也许是数据可视化最基本和最常用的用法，但这并不意味着它没有价值。原因是大多数数据都涉及时间因素。因此，许多数据分析的第一步是查看数据如何随时间而变化。

（2）确定频率

频率也是数据可视化的相当基本的用法，因为它也适用于涉及时间的数据。如果涉及时间，则应该确定相关事件随时间发生的频率是合乎逻辑的。

（3）确定关系（相关性）

识别关联是数据可视化极有价值的用途。如果没有可视化，则极难确定两个变量之间的关系。了解数据中的关系非常重要。这是数据可视化在数据分析中的价值的一个很好例子。

（4）检查人际网络

在市场研究中可以看到使用数据可视化检查人际网络的示例。市场营销专业人员需要知道他们的消息针对哪些受众，因此他们需要分析整个市场以识别受众群体、集群之间的桥梁、集群内的意见领袖以及异常值。

（5）制订计划

为一个复杂的项目制订计划或时间表时，事情可能会变得混乱。甘特图通过清楚地说明项目中的每个任务以及完成所需的时间来解决该问题。

（6）分析价值和风险

确定诸如价值和风险之类的复杂指标需要考虑许多不同的变量，而使用普通电子表格几乎是无法准确看到的。数据可视化可以像对公式进行颜色编码一样简单，以显示哪些机会有价值，哪些机会具有风险。

**4. 数据可视化图表的类型**

现在，我们了解了如何使用数据可视化，以及让我们应用不同类型的数据可视化。有许多工具可用来帮助创建数据可视化，有些是手动的，有些是自动化的，但是无论哪种方式，它们都应允许你进行以下任何类型的可视化处理。

（1）折线图

折线图说明了随着时间的变化而发生的情况。x 轴通常是一段时间，而 y 轴是数量。因此，这可以说明公司按月份细分的年度销售额，或者工厂过去一周内每天生产多少台设备。

（2）面积图

面积图是对折线图的改编，其中线下的区域被填充以强调其重要性。每条线下方区域的填充颜色应有些透明，以便可以识别重叠区域。

（3）条形图

条形图说明了随着时间的变化而发生的情况。但是，如果有多个变量，则条形图可以更加容易地比较每个变量在每个时间点的数据。例如，条形图可以比较公司从去年到今年的销售额。

（4）直方图

直方图看起来像条形图，但它测量的是频率而不是随时间变化的趋势。直方图的 x 轴列出了变量的"区间"或间隔，而 y 轴是频率，因此每个条形表示该区间的频率。例如，你可以测量调查问题的每个答案的频率。区间将是答案"不令人满意""中立"和"令人满意"。这将告诉你每个答案有多少人。

（5）散点图

散点图用于查找相关性。散点图上的每个点都表示"当 x 等于某值时，则 y 等于某值"。这样，如果这些点具有某种趋势，则它们之间存在关系；如果该图确实是零散的，没有任何趋势，则这些变量完全不会相互影响。

（6）气泡图

气泡图是散点图的一种改编，其中每个点都显示为气泡，其区域除在轴上的位置外还具有含义。与气泡图相关的痛点是由于轴内空间有限，气泡尺寸受到了限制。因此，这种类型的可视化并不适合有效地展示所有数据。

（7）饼图

饼图是说明百分比的最佳选择，因为它将每个元素都显示为整体的一部分。因此，如果你的数据说明了百分比细分，则饼图将清楚地按适当的比例显示各个部分。

（8）仪表盘图

仪表盘图可用于说明间隔之间的距离。它可以表示为圆形钟表或类似于液体温度计的管型表。多个仪表盘图可以彼此相邻显示，以说明多个间隔之间的差异。

（9）地图

企业中处理的许多数据都有一个位置元素，这使得在地图上显示变得容易。地图可视化的一个示例是映射每个州的客户购买数量。在此示例中，每个州都用颜色标记，购买次数较少的州的颜色更浅一些，而购买次数较多的州的颜色则更深一些。位置信息对于企业领导者了解企业情况也非常有价值，这就使地图成为重要的数据可视化图表类型。

（10）热图

热图基本上是一种颜色编码的矩阵。使用公式为矩阵的每个单元着色，以表示该单元的相对值或风险。通常，热图的颜色范围从绿色到红色，其中绿色表示效果更好，红色效果表示更差。这种类型的可视化很有用，因为颜色比数字更容易解释。

（11）框架图

框架图基本上是树状图，清楚地显示了层次关系结构。框架图由分支组成，每个分支都有更多的分支与之连接，图的每个级别都由越来越多的分支组成。

**5. 结论**

有效的数据可视化是数据分析至关重要的最后一步。没有它，重要的见解和信息可能会丢失。

# Text B

## 10 Best Data Visualization Tools

扫码听课文

In this era of data, it is very important to understand the data to obtain some actionable insights. And data visualization is a very important part of understanding the hidden patterns and layers in the data. What sounds more interesting to you? A beautiful and descriptive bar chart or a boring spreadsheet telling the same information?

Of course it is the bar chart because humans are visual creatures. Data visualization charts like bar charts, scatter plots, line charts, geographical maps, etc. are extremely important. They tell you information just by looking at them whereas normally you would have to read spreadsheets or text reports to understand the data.

There are some data visualization tools which are very popular as they allow analysts and statisticians to create visual data models easily according to their specifications by conveniently providing an interface, database connections and machine learning tools all in one place. The following are the 10 best data visualization tools.

## 1. Tableau

Tableau is a data visualization tool that can be used by data analysts, scientists, statisticians, etc. to visualize the data and get a clear opinion based on the data analysis. Tableau is very famous as it can take in data and produce the required data visualization output in a very short time. And it can do this while providing the highest level of security with a guarantee to handle security issues as soon as they arise or are found by users.

Tableau also allows its users to prepare, clean and format their data and then create data visualizations to obtain actionable insights that can be shared with other users. Tableau is available for the individual data analyst or at scale for business teams and organizations.

## 2. Looker

Looker is a data visualization tool that can go in-depth in the data and analyze it to obtain useful insights. It provides real-time dashboards of the data for more in-depth analysis so that businesses can make instant decisions based on the data visualizations obtained. Looker also provides connections with Redshift, Snowflake, BigQuery, as well as more than 50 SQL supported dialects so you can connect to multiple databases without any issues.

Looker data visualizations can be shared with anyone using any particular tool. Also, you can export these files in any format immediately. It also provides customer support wherein you can ask any question and it shall be answered.

## 3. Zoho Analytics

Zoho Analytics is a business intelligence and data analytics software that can help you create wonderful looking data visualizations based on your data in a few minutes. You can obtain data from multiple sources and mesh it together to create multidimensional data visualizations that allow you to view your business data across departments. In case you have any questions, you can use Zia which is a smart assistant created using artificial intelligence, machine learning and natural language processing.

Zoho Analytics allows you to share or publish your reports with your colleagues and add comments or engage in conversations as required. You can export Zoho Analytics files in any format such as spreadsheet, Word, Excel, PPT, PDF, etc.

## 4. Sisense

Sisense is a business intelligence-based data visualization system and it provides various tools that allow data analysts to simplify complex data and obtain insights for their organization and outsiders. Sisense believes that eventually every company will be a data-driven company and every product will be related to data in some way. Therefore it tries its best to provide various data analytics tools to business teams and data analytics so that they can help make their companies the data-driven companies of the future.

It is very easy to set up and learn Sisense. It can be easily installed within a minute and data analysts can get their work done and obtain results instantly. Sisense also allows its users to export their files in multiple formats such as PPT, Excel, Word, PDF, etc. Sisense also provides full-time customer support services whenever users face any issues.

**5. IBM Cognos Analytics**

IBM Cognos Analytics is an artificial intelligence-based business intelligence platform that supports data analytics among other things. You can visualize as well as analyze your data and share actionable insights with anyone in your organization. Even if you have limited or no knowledge about data analytics, you can use IBM Cognos Analytics easily as it interprets the data for you and presents you with actionable insights in plain language.

You can also share your data with multiple users if you want on the cloud and share visuals over email or Slack. You can also import data from various sources like spreadsheets, cloud, CSV files or on-premises databases and combine related data sources into a single data module.

**6. Qlik Sense**

Qlik Sense is a data visualization platform that helps companies to become data-driven enterprises by providing an associative data analytics engine, sophisticated artificial intelligence system and scalable multi-cloud architecture that allows you to deploy any combination of SaaS, on-premises or a private cloud.

You can easily combine, load, visualize and explore your data on Qlik Sense, no matter what its size is. All the data charts, tables and other visualizations are interactive and instantly update themselves according to the current data context. The Qlik Sense AI can even provide you with data insights and help you create analytics using just drag and drop.

**7. Domo**

Domo is a business intelligence model that contains multiple data visualization tools which provide a consolidated platform where you can perform data analysis and then create interactive data visualizations that allow other people to easily understand your data conclusions. You can combine cards, text and images in the Domo dashboard so that you can guide other people through the data while telling a data story as they go.

In case of any doubts, you can use their pre-built dashboards to obtain quick insights from the data.

**8. Microsoft Power BI**

Microsoft Power BI is a data visualization platform focused on creating a data-driven business intelligence culture in all companies today. To fulfill this, it offers self-service analytics tools that can be used to analyze, aggregate, and share the data in a meaningful fashion.

Microsoft Power BI offers hundreds of data visualizations to its customers along with built-in artificial intelligence capabilities and Excel integration facilities. It also provides you with multiple support systems such as FAQs, forums and also live chat support with the staff.

**9. Klipfolio**

Klipfolio is one of the best data visualization tools. You can access your data from hundreds of different data sources like spreadsheets, databases, files and web services applications by using connectors. Klipfolio also allows you to create custom drag-and-drop data visualizations wherein you can choose from different options like charts, graphs, scatter plots, etc.

**10. SAP Analytics Cloud**

SAP Analytics Cloud uses business intelligence and data analytics capabilities to help you evaluate your data and create visualizations in order to predict business outcomes. It also provides you with the latest modeling tools that help you by alerting you of possible errors in the data and categorizing different data measures and dimensions. SAP Analytics Cloud also suggests smart transformations to the data that lead to enhanced visualizations.

In case you have any doubts or business questions related to data visualization, SAP Analytics Cloud provides you with complete customer satisfaction by handling your queries using conversational artificial intelligence and natural language technology.

## ✎ New Words

| | | |
|---|---|---|
| obtain | [əb'teɪn] | v.得到；存在 |
| actionable | ['ækʃənəbl] | adj.可行动的 |
| insight | ['ɪnsaɪt] | n.见解；洞察力；领悟 |
| conveniently | [kən'viːniəntli] | adv.方便地，便利地 |
| opinion | [ə'pɪnjən] | n.意见，主张 |
| guarantee | [,gærən'tiː] | v.担保；确保<br>n.保证；保修单 |
| dialect | ['daɪəlekt] | n.方言，专业用语 |
| export | [ɪk'spɔːt] | v.（计算机）导出 |
| mesh | [meʃ] | v.（使）吻合；相配，匹配；紧密配合 |
| conversation | [,kɒnvə'seɪʃn] | n.交谈，会话；人机对话 |
| simplify | ['sɪmplɪfaɪ] | vt.简化 |
| consolidate | [kən'sɒlɪdeɪt] | v.统一；合并；联合 |
| pre-built | [priː'bɪlt] | adj.预建的 |
| forum | ['fɔːrəm] | n.论坛 |
| connector | [kə'nektə] | n.连接器 |
| evaluate | [ɪ'væljueɪt] | v.估计 |
| measure | ['meʒə] | v.衡量；测量；估量<br>n.尺度；度量单位 |

## ✎ Phrases

| | |
|---|---|
| text report | 文本报告 |
| visual data model | 可视数据模型 |
| data visualization tool | 数据可视化工具 |

| be shared with | 与……共享 |
| data analytics software | 数据分析软件 |
| multidimensional data visualization | 多维数据可视化 |
| smart assistant | 智能助理 |
| data-driven company | 数据驱动公司 |
| data module | 数据模块 |
| drag and drop | 拖放 |
| self-service analytics tool | 自助分析工具 |
| web services application | 网络服务应用程序 |

## ✎ Abbreviations

| CSV(Comma-Separated Values) | 逗号分隔值，也称为字符分隔值 |
| FAQ (Frequently Asked Questions) | 常见问题，经常问到的问题 |

## Reference Translation

## 十种最佳数据可视化工具

在这个数据时代，了解数据以获得一些可行的见解非常重要。数据可视化是了解数据中隐藏模式和层的非常重要的部分。什么听起来更有趣呢？漂亮的、描述性的条形图与无聊的电子表格告诉你的信息能一样吗？

当然是条形图，因为人类是视觉生物。数据可视化图表（如条形图、散点图、折线图、地理地图等）非常重要。仅通过查看这些图表你就能获得信息，而通常你必须阅读电子表格或文本报告才能理解数据。

有一些数据可视化工具非常受欢迎，因为这些工具集中提供了界面、数据库连接和机器学习工具，它们让分析人员和统计人员可以轻松地按照规范创建可视数据模型。以下是十种最佳数据可视化工具。

### 1. Tableau

Tableau 是一种数据可视化工具，数据分析师、科学家、统计学家等可使用它实现数据可视化，并基于数据分析获得明确的意见。Tableau 非常有名，因为它可以接收数据并在很短的时间内输出所需的数据可视化。并且同时提供最高级别的安全性，还保证在出现安全问题或用户发现问题后立即对其进行处理。

Tableau 还允许其用户准备、清理和格式化他们的数据，然后创建数据可视化，以获取可与其他用户共享的可行性见解。单个数据分析师可以使用 Tableau，业务团队和组织也可使用。

### 2. Looker

Looker 是一个数据可视化工具，可以深入到数据中并对其进行分析以获得有用的见解。它提供了数据的实时仪表板，以进行更深入的分析，这样企业就可以根据所获得的数据可视化来做出即时决策。Looker 还提供与 Redshift、Snowflake、BigQuery 以及超过 50 种 SQL 支持的方言的连接，因此你可以连接到多个数据库。

Looker 数据可视化可以与使用任何特定工具的任何人共享。另外，你可以立即以任何格式导出这些文件。它还提供客户支持，你可以在其中提出任何问题，并且会得到答复。

## 3. Zoho Analytics

Zoho Analytics 是一款商业智能和数据分析软件，可以在几分钟内基于你的数据帮助创建外观精美的数据可视化。你可以从多个来源获取数据并将其网格化以创建多维数据可视化，以便查看跨部门的业务数据。如有任何疑问，可以使用 Zia，它是使用人工智能、机器学习和自然语言处理创建的智能助手。

Zoho Analytics 允许你与同事共享或发布报告，并根据需要添加评论或进行对话。你可以以任何格式导出 Zoho Analytics 文件，如电子表格、Word、Excel、PPT、PDF 等。

## 4. Sisense

Sisense 是一个基于商业智能的数据可视化系统，它提供了各种工具，这些工具使数据分析师可以简化复杂数据并获取对其组织和外部人员的见解。Sisense 相信，每个公司最终都将成为数据驱动的公司，并且每个产品都将以某种方式与数据相关。因此，它竭尽全力为业务团队和数据分析提供各种数据分析工具，以便帮助他们的公司成为未来数据驱动的公司。

设置和学习 Sisense 非常容易。它可以在一分钟内轻松安装，数据分析师可以立即完成工作并获得结果。Sisense 还允许其用户以多种格式（如 PPT、Excel、Word、PDF 等）导出文件。Sisense 还可以在用户遇到任何问题时提供全职客户支持服务。

## 5. IBM Cognos Analytics

IBM Cognos Analytics 是一个基于人工智能的商业智能平台，除其他功能外还支持数据分析。你可以可视化并分析数据，并与组织中的任何人共享可行的见解。即使你对数据分析的了解有限或根本不了解，你也可以轻松使用 IBM Cognos Analytics，因为它可以为你解释数据，并以通俗易懂的语言为你提供可行的见解。

你还可以在云上与多个用户共享数据，并通过电子邮件或 Slack 共享视觉效果。你还可以从各种数据源（如电子表格、云、CSV 文件或本地数据库）导入数据，并将相关数据源合并到一个数据模块中。

## 6. Qlik Sense

Qlik Sense 是一个数据可视化平台，它通过提供关联的数据分析引擎、先进的人工智能系统和可扩展的多云体系结构来帮助公司成为数据驱动型企业，该体系结构允许你部署 SaaS、本地或私有云的任何组合。

无论大小如何，你都可以在 Qlik Sense 上轻松组合、加载、可视化和浏览数据。所有数据图表、表格和其他可视化视图都是交互式的，并根据当前数据情况即时更新。Qlik Sense AI 甚至可以为你提供数据洞察力，并仅通过拖放即可帮助你创建分析。

## 7. Domo

Domo 是一个商业智能模型，其中包含多个数据可视化工具，这些工具提供了一个整合的平台，你可以在其中执行数据分析，然后创建交互式数据可视化，以使其他人可以轻松地理解你的数据结论。你可以在 Domo 仪表板中组合卡片、文本和图像，以便在给其他人讲述数据故事的同时引导他们浏览数据。

如有任何疑问，你可以使用其预先构建的仪表板从数据中获得快速的见解。

**8. Microsoft Power BI**

Microsoft Power BI 是一个数据可视化平台，致力于在当今所有公司中创建一种数据驱动的商业智能文化。为此，它提供了自助服务分析工具，可用于以有意义的方式分析、汇总和共享数据。

Microsoft Power BI 以及内置的人工智能功能和 Excel 集成工具为客户提供了数百种数据可视化方法。它还提供多种支持系统，如常见问题解答、论坛以及与员工的实时聊天支持。

**9. Klipfolio**

Klipfolio 是最好的数据可视化工具之一。你可以使用连接器从数百个不同的数据源（如电子表格、数据库、文件和 Web 服务应用程序）访问数据。Klipfolio 还允许你创建自定义的拖放数据可视化，你可以在其中选择不同的选项，如图表、图形、散点图等。

**10. SAP Analytics Cloud**

SAP Analytics Cloud 使用商业智能和数据分析功能来帮助你评估数据并创建可视化以预测业务成果。它还为你提供了最新的建模工具，提醒你数据中可能存在的错误并对不同的数据度量和维度进行分类。SAP Analytics Cloud 还建议对数据进行智能转换，以增强可视化效果。

如果你对数据可视化有任何疑问或业务问题，SAP Analytics Cloud 通过使用对话式人工智能和自然语言技术来处理你的查询，可以使你完全满意。

# Exercises

**[Ex. 1]** 根据 **Text A** 回答以下问题。

1. What is data visualization?

2. Why do we need data visualization?

3. What is frequency?

4. What do marketing professionals do to know which audiences to target with their message?

5. What happens when planning out a schedule or timeline for a complex project? What kind of chart addresses that issue and how?

6. What is an area chart?

7. What is a bubble chart?

8. Why is a pie chart the best option for illustrating percentages?

9. What is a heat map? Why is this type of visualization helpful?

10. What does a frame diagram consist of?

[Ex. 2] 根据 Text B 回答以下问题。

1. What is Tableau?

2. What does Looker provide?

3. What is Zoho Analytics?

4. What does Sisense believe?

5. Why can you use IBM Cognos Analytics easily even if you have limited or no knowledge about data analytics?

6. What is Qlik Sense?

7. What is Domo?

8. What does Microsoft Power BI offer?

9. What can you do with Klipfolio?

10. What does SAP Analytics Cloud use business intelligence and data analytics capabilities to help you to do? What does it also provide you with?

[Ex. 3] 词汇英译中

1. area chart      1. _____

2. batch      2. _____

3. bar chart      3. _____

4. chart      4. _____

5. bubble chart      5. _____

6. formula      6. _____

7. frame diagram      7. _____

8. histogram      8. _____

9. heat map      9. _____

10. manual      10. _____

**[Ex. 4]** 词汇中译英

1. 饼图                            1. _____
2. 矩阵；模型                      2. _____
3. 数据模块                        3. _____
4. 阴影；色度                      4. _____
5. 拖放                            5. _____
6. 可视化，形象化；虚拟化          6. _____
7. 智能助理                        7. _____
8. 估计                            8. _____
9. 可视数据模型                    9. _____
10. 见解；洞察力；领悟            10. _____

**[Ex. 5]** 短文翻译

### What Is Visual Data Analysis

Sometimes, data can be overwhelming. There's too much of it, too little time to comprehend it, or you simply can't see the data you have available at your disposal. If so, visual data analysis can help you make sense of it all, by combining data analytics and data visualization techniques. Data analytics alone can be powerful. However, it can be difficult to see the big picture or how one set of data relates to another. Visualization tools by themselves may make static mashups and presentations of data easy to grasp. Visual data analysis brings you the best of both worlds.

#### 1. Components of a visual data analysis solution

A visual data analysis solution will have an interface, often an interactive dashboard on a screen, for users to select sources of data and choices for displaying the data. Data display options may range from basic line, bar and pie charts to more sophisticated gauge indicators, scatter charts and tree maps. Ideally, the requirements for the solution are as follows.

- Need only little or preferably no coding.
- Make it simple to locate and bring in data from different sources.
- Offer graphics that are easy to customize.
- Make it easy to drill down to the underlying data at any level of detail.
- Make it possible to combine multiple views for at-a-glance, overall understanding.

#### 2. What is visual data analysis best used for

Visual data analysis makes it easier for human beings to understand data, broad relationships and patterns can be brought out. An interactive dashboard can also be a great tool to explain the story about the data to others, and for answering their questions about the data and possible insights as they think of them. There is another very useful side to visual data analysis too. Visualization helps you quickly narrow your search for information of interest. You can then apply data analytics algorithms to the datasets for a thorough analysis and report. Visualization also lets you apply common sense to check intermediate results. You catch any mistakes earlier. By toggling between data analytics and visualization, you can home in faster on datasets of interest and insights of value.

# Reading Material

## 25 Tips for Data Visualization Design

### 1. Story

To start, let's cover a few general things to keep in mind. Remember that every data visualization design choice you make should enhance your reader's experience, not yours.

1) Choose the chart that tells the story. There may be more than one way to visualize the data accurately. In this case, consider what you're trying to achieve, the message[1] you're communicating, who you're trying to reach, etc.

2) Remove anything that doesn't support the story. That doesn't mean you kill half your data points. But be mindful of things like chart junk, extra copy, unnecessary illustrations, drop shadows, ornamentations, etc. The great thing about data visualization is that design can help do the heavy lifting to enhance and communicate the story.

3) Design for comprehension[2]. Once you have your visualization created, take a step back and consider what simple elements might be added, tweaked or removed to make the data easier for the reader to understand. You might add a trend line to a line chart, or you might realize you have too many slices in your pie chart (use 6 max). These subtle tweaks make a huge difference.

### 2. Comparison

Data visualization makes comparison a lot easier, letting you actually "see" how two different data sets stack up[3] to each other. But just putting two charts side by side doesn't necessarily accomplish that. In fact, it can make it more confusing[4].

1) Include a zero baseline[5] if possible. Although a line chart does not have to start at a zero baseline, it should be included if it gives more context for comparison and if relatively small fluctuations[6] in data are meaningful (e.g., in stock market data).

2) Always choose the most efficient visualization. You want visual consistency[7] so that the reader can compare at a glance. This might mean you use stacked bar charts, a grouped bar chart or a line chart. Whatever you choose, don't overwhelm by making the reader work to compare too many things.

3) Watch your placement. You may have two nice stacked bar charts that are meant to let your reader compare points, but if they're placed too far apart to "get" the comparison, you've already lost.

---

1　message ['mesɪdʒ] *n.*信息　*v.*给……发消息

2　comprehension [ˌkɒmprɪ'henʃn] *n.*理解；充分了解

3　stack up: 叠加；把……堆在……

4　confusing [kən'fjuːzɪŋ] *adj.*令人困惑的；混乱的；混淆的

5　baseline ['beɪslaɪn] *n.*基线

6　fluctuation [ˌflʌktʃu'eɪʃn] *n.*波动，涨落，起伏

7　consistency [kən'sɪstənsi] *n.*连贯性，一致性

4) Tell the whole story. Maybe you had a 30% sales increase in Q4. Exciting! But what's more exciting? Showing that you've actually had a 100% sales increase since Q1.

## 3. Copy

Data is about numbers, certainly, but it is generally used in conjunction with copy to help provide context for the point at hand.

1) Don't over explain. If the copy already mentions a fact, the subhead[8], callout[9] and chart header don't have to reiterate it.

2) Keep chart and graph headers simple and to the point. There's no need to get verbose[10]. Keep any descriptive[11] text above the chart brief and directly related to the chart underneath.

3) Use callouts wisely. Callouts are not there to fill space. They should be used intentionally to highlight relevant information or provide additional context.

4) Don't use distracting fonts or elements. Sometimes you do need to emphasize a point. If so, only use bold or italic text to emphasize a point — and don't use them both at the same time.

## 4. Color

Color is a great tool when used well. When used poorly, it can not just distract but misdirect the reader. Use it wisely in your data visualization design.

1) Use a single color to represent the same type of data. If you are depicting sales month by month on a bar chart, use a single color. But if you are comparing last year's sales to this year's sales in a grouped chart, you should use a different color for each year. You can also use an accent color to highlight a significant data point.

2) Watch out for positive and negative numbers[12]. Don't use red for positive numbers or green for negative numbers. Those color associations are so strong it will automatically flip[13] the meaning in the viewer's mind.

3) Make sure there is sufficient contrast between colors. If colors are too similar (light gray, light and light gray), it can be hard to tell the difference.

4) Avoid patterns. Stripes and polka dots sound fun, but they can be incredibly distracting. If you want to differentiate on a map, use different saturations of the same color.

5) Select colors appropriately. Some colors stand out more than others, giving unnecessary weight to that data. Instead, you can use a single color with varying shade or a spectrum between two analogous colors to show intensity. Remember to intuitively code color intensity[14] according to values as well.

6) Don't use more than 6 colors in a single layout.

---

8  subhead ['sʌbhed] *n.*小标题，副标题

9  callout [ˌkɒl'aut] *n.*插图的编号

10  verbose [vɜ:'bəus] *adj.*冗长的，啰唆的，累赘的

11  descriptive [dɪ'skrɪptɪv] *adj.*描写的，叙述的

12  negative number: 负数

13  flip [flɪp] *vt.*快速翻转

14  intensity [ɪn'tensəti] *n.*强度

## 5. Labeling

Labeling can be a minefield[15]. Readers rely on labels to interpret data, but too many or too few can interfere.

1) Double check that everything is labeled. Make sure everything that needs a label has one, and that there are no doubles or typos.

2) Make sure labels are visible. All labels should be unobstructed[16] and easily identified with the corresponding data point.

3) Label the lines directly. If possible, include data labels with your data points. This lets readers quickly identify lines and corresponding labels so they don't have to go hunting for a legend[17] or similar point.

4) Don't over label. If the precise value of a data point is important to tell your story, then include data labels to enhance comprehension. If the precise values are not important to tell your story, leave the data labels out.

5) Don't set your type at an angle. If your axis labels are too crowded[18], consider removing every other label on an axis to allow the text to fit comfortably.

## 6. Ordering

Data visualization is meant to help make sense. Random[19] patterns that are difficult to interpret are frustrating and detrimental to what you're trying to communicate.

1) Order data intuitively. There should be a logical hierarchy. Order categories alphabetically, sequentially or by value.

2) Order consistently. The ordering of items in your legend should mimic[20] the order of your chart.

3) Order evenly[21]. Use natural increments on your axes (0, 5, 10, 15, 20) instead of awkward or uneven increments (0, 3, 5, 16, 50).

---

15　minefield ['maɪnfiːld] n.雷区

16　unobstructed [ˌʌnəb'strʌktɪd] adj.不被阻塞的，没有障碍的

17　legend ['ledʒənd] n.图例

18　crowded ['kraʊdɪd] adj.拥挤的；充满的

19　random ['rændəm] adj.随机的；任意的

20　mimic ['mɪmɪk] vt.模仿，模拟

21　evenly ['iːvnli] adv.均匀地；均等地

# Unit 9

## Text A

## Big Data Cloud

**1. The fusion of big data with cloud technology**

When big data computing takes place in the clouds it is known as "Big Data Clouds". Their purpose is to build an integrated infrastructure that is suitable for quick analytics and deployment of an elastically scalable infrastructure. Cloud technology is used to derive quantum-leap advantages inherent in big data.

扫码听课文

(1) Features of big data clouds

The features of big data clouds are as follows.

1) Large-scale distributed computing: A wide range of computing facilities intended for distributed architectures.

2) Data storage: These are all the scalable storage facilities and data services that work seamlessly.

3) Metadata-based data access: The insight generated from the data is used instead of using path and filenames.

4) Distributed virtual file system: Files are arranged in a hierarchical structure, where the nodes are the directories.

5) High-performance data and computation: The data, as well as computations, is driven and enhanced by performance.

6) Analytics service provider: It helps to develop, deploy and use analytics.

7) Multi-dimension data: It provides the needed support for various data tools and types to facilitate their processing.

8) Analytics service provider: It enables the development, deployment and use of analytics.

9) High availability of data and computing: Duplication mechanism of both data and computation is carried out to make them always available.

10) Integrated platform: It gives the needed platform for fast analytics and deployment of scalable architecture.

(2) Types of big data clouds

Big data clouds are classified into four different types on the basis of their usage.

1) Public big data cloud: In this cloud, resources are offered as a pay-as-go computing model

and stretch over a large-scale organization. This cloud is elastically scalable in terms of architecture. For example Google cloud platform of big data computing, Windows Azure HDInsight, Amazon big data computing in clouds and RackSpace Cloudera Hadoop.

2) Private big data cloud: This is the cloud within an organization with the aim of providing privacy and greater resource control through virtualized infrastructure.

3) Hybrid big data cloud: This cloud espouses the characteristics and functionalities of both private and public clouds. Its purpose is to attain network configuration and latency, scalability, high availability, data sovereignty, disaster recovery, compliance and workload portability. During the period of heavy workloads, its deployment facilitates the migration of private workloads to public architecture.

4) Big data access networks and computing platform: This is an integrated platform designed for data, analytics and computing. This platform provides services by multiple distinct providers.

(3) Elements and services

The fusion of big data and cloud technology gives rise to many elements and services.

1) Cloud mounting services or mount cloud storage: It is a system designed to mount data to many web servers and cloud storage. For example Amazon, Google drive.

2) Big data cloud: It is an infrastructure designed to manage different data sources like data management, data access, security, schedulers, programming models, etc. The big data cloud infrastructure has many tools like streaming, web services and APIs for collecting data from different sources such as Google data, social networking, relational data stores, etc.

3) Cloud streaming: It is the act of transferring multimedia data such as video and audio in a way that makes them readily available in continuous mode over social media infrastructure.

4) Social media: These are platforms that enable the gathering of a multitude of information online. It is the major source of big data utilizing the services of cloud technology. Examples are Facebook, LinkedIn, Twitter and YouTube.

5) HTTP, REST services: These services are designed to develop APIs for web-based applications that are lightweight, scalable and maintainable. For example USGS Geo Cloud, Google Earth.

6) Big data computing platform: It is designed to create modules required to manage different data sources like data-intensive programming models, data security, compute and data-aware scheduling, appliances, distributed file system and analytics.

7) Computing cloud: This platform is developed to offer the required computing infrastructure like physical and virtual compute as well as storage from private, public and hybrid clouds.

## 2. Big data cloud reference architecture

The cloud architecture for big data is efficient to manage complicated computing scalability, storage and networking infrastructure. The infrastructure as service providers mainly deals with servers, networks, in addition to storage applications and offers facilities such as virtualization, basic monitoring and safety, operating system, server in a data center and storage services. The four layers of large data cloud architecture are discussed below.

(1) Big data analytics—Software as a Service (SaaS)

The analytics of big data offered as service gives users the capability to quickly work on analytics without spending on infrastructure and paying for the facilities used. The functions of this layer are as follows.

- Arrangement of software applications repository.
- Software programs deployment on the infrastructure.
- Result delivery to the users.

(2) Big data analytics—Platform as a Service (PaaS )

This is the second layer of the architecture. It is the core layer that provides platform-related services to work with stored big data and computing. Data management tools, schedulers and programming environments for data-intensive and data processing tasks, which are considered as middleware management tools, reside in this region. This layer is responsible for providing software development kits and tools necessary for analytics.

(3) Big Data Fabric (BDF)

This is the fabric layer of big data, responsible for addressing tools and APIs that support the storage of data, data computation and access to different application services. This layer comprises APIs and interoperable protocol designed to connect the specified multiple cloud infrastructural standards.

(4) Cloud Infrastructure (CI)

The cloud infrastructure is responsible for handling the infrastructure for data storage and computation as services. The services offered by CI layer are as follows.

- To create large-scale elastic infrastructure for big data storage, which is capable of on-demand deployment.
- To set up dynamic virtual machines.
- To generates on-demand storage facilities that relate to big data management for file, block and object-based.
- To enable seamless passage of data across the storage repositories.
- To create virtual machines and to mount the file system with the compute node.

## 3. Benefits of big data analysis in cloud

(1) Improved analysis

With the advancement of cloud technology, big data analysis has become more improved, causing better results. Hence, companies prefer to perform big data analysis in the cloud. Moreover, Cloud helps to integrate data from numerous sources.

(2) Simplified infrastructure

Big data analysis is a tremendous strenuous job on infrastructure as the data comes in large volumes with varying speeds and types which traditional infrastructures usually cannot keep up with. As cloud computing provides flexible infrastructure, which we can scale according to the needs at the time, it is easy to manage workloads.

(3) Lowering the cost

Both big data and cloud technology delivers value to organizations by reducing the ownership. Cloud enables customers for big data processing without large-scale big data resources. Hence, both big data and cloud technology are driving the cost down for enterprise purposes and bringing value to the enterprise.

(4) Security and privacy

Data security and privacy are two major concerns when dealing with enterprise data. On the one hand, security becomes a primary concern when your application is hosted on a cloud platform due to its open environment and limited user control. On the other hand, being an open source application, big data solution like Hadoop uses a lot of third-party services and infrastructure. Hence, nowadays system integrators bring in private cloud solution that is elastic and scalable. Furthermore, it also leverages scalable distributed processing.

Besides that, cloud data is stored and processed in a central location commonly known as cloud storage server. Along with it the service provider and the customer sign a Service Level Agreement (SLA) to gain the trust between them. If required, the provider also leverages required advanced level of security control.

There are rules associated with service level agreements for protecting data, capacity, scalability, security, privacy, availability of data storage and data growth.

In many organizations, big data analytics is utilized to detect and prevent advanced threats and malicious hackers.

(5) Virtualization

Infrastructure plays a crucial role to support any application. Virtualization technology is the ideal platform for big data. Virtualized big data applications like Hadoop provide multiple benefits which are not accessible on physical infrastructure, but it simplifies big data management. Big data and cloud computing point to the convergence of various technologies and trends that makes IT infrastructure and related applications more dynamic and more modular. Hence, big data and cloud computing projects rely heavily on virtualization.

## New Words

| | | |
|---|---|---|
| elastically | [ɪˈlæstɪkli] | adv.有弹性地，伸缩自如地 |
| quantum-leap | [ˌkwɒntəmˈliːp] | n.巨大突破 |
| inherent | [ɪnˈhɪərənt] | adj.固有的，内在的 |
| filename | [ˈfaɪlneɪm] | n.文件名 |
| directory | [dəˈrektəri] | n.（计算机文件或程序的）目录 |
| high-performance | [ˌhaɪ pəˈfɔːməns] | adj.高性能的 |
| deploy | [dɪˈplɔɪ] | v.部署 |
| multi-dimension | [ˈmʌltɪ daɪˈmenʃn] | n.多维 |
| duplication | [ˌdjuːplɪˈkeɪʃn] | n.复制；重复；成倍 |
| stretch | [stretʃ] | v.伸展；绵延，延续 |
| espouse | [ɪˈspaʊz] | vt.支持；拥护；赞助 |

| portability | [ˌpɔːtəˈbɪləti] | n.可移植性，可迁移性 |
| migration | [maɪˈɡreɪʃn] | n.迁移 |
| distinct | [dɪˈstɪŋkt] | adj.明显的，清楚的 |
| data-intensive | [ˈdeɪtə ɪnˈtensɪv] | adj.数据密集的 |
| middleware | [ˈmɪdlweə] | n.中间设备，中间件 |
| comprise | [kəmˈpraɪz] | vt.包含，包括；由……组成，由……构成 |
| on-demand | [ˌɒn dɪˈmɑːnd] | adj.按需的 |
| block | [blɒk] | n.块 |
| object-based | [ˈɒbdʒɪkt beɪst] | adj.基于对象的 |
| numerous | [ˈnjuːmərəs] | adj.很多的，许多的 |
| tremendous | [trəˈmendəs] | adj.极大的，巨大的；可怕的，惊人的 |
| strenuous | [ˈstrenjuəs] | adj.费力的，用力的 |
| concern | [kənˈsɜːn] | n.令人担心的事；关心（的事） v.影响；涉及；使担忧 |
| integrator | [ˈɪntɪɡreɪtə] | n.综合者；积分器（仪、机、元件、电路、装置） |
| leverage | [ˈliːvərɪdʒ] | v.利用；发挥杠杆作用 n.杠杆作用；优势，力量 |
| convergence | [kənˈvɜːdʒəns] | n.集合；会聚 |

## ✍ Phrases

| take place | 发生 |
| big data cloud | 大数据云 |
| be suitable for | 适合……的 |
| large-scale distributed computing | 大规模分布式计算 |
| distributed virtual file system | 分布式虚拟文件系统 |
| hierarchical structure | 层次结构，分级结构 |
| be classified into | 分成，分为 |
| computing model | 计算模型 |
| virtualized infrastructure | 虚拟化基础设施 |
| network configuration | 网络配置 |
| data sovereignty | 数据主权 |
| disaster recovery | 灾难恢复 |
| workload portability | 工作负载可移植性 |
| distributed file system | 分布式文件系统 |
| virtual compute | 虚拟计算 |
| computation as service | 计算即服务 |

| dynamic virtual machine | 动态虚拟机 |
|---|---|
| storage repository | 存储库 |
| virtual machine | 虚拟机 |
| third-party service | 第三方服务 |
| physical infrastructure | 物理基础设施 |

## ✎ Abbreviations

| REST (REpresentational State Transfer) | 表现层状态转移，表述性状态传递 |
|---|---|
| PaaS (Platform as a Service) | 平台即服务 |
| BDF (Big Data Fabric) | 大数据结构 |
| CI (Cloud Infrastructure) | 云基础设施 |

# Reference Translation

# 大 数 据 云

## 1. 大数据与云技术的融合

当大数据计算在云中进行时，被称为"大数据云"。它的目的是建立集成基础架构，以适应快速分析和部署弹性可扩展基础架构。云技术用于获得大数据固有的量子飞跃优势。

（1）大数据云的特征

大数据云的特征如下：

1）大规模分布式计算：用于分布式体系结构的各种计算工具。

2）数据存储：这些都是无缝可扩展存储工具和数据服务。

3）基于元数据的数据访问：使用从数据生成的见解，而不是使用路径和文件名。

4）分布式虚拟文件系统：文件以分层结构排列，其中节点是目录。

5）高性能数据和计算：数据和计算均由性能来驱动与增强。

6）分析服务提供商：它有助于开发、部署和使用分析。

7）多维数据：它为各种数据工具和类型提供了所需的支持，以促进其处理。

8）分析服务提供商：它支持分析的开发、部署和使用。

9）数据和计算的高可用性：执行数据和计算的复制机制以使其始终可用。

10）集成平台：它为快速分析和部署可扩展的体系结构提供了所需的平台。

（2）大数据云的类型

大数据云根据其用途分为四种不同类型。

1）公共大数据云：在这种云中，资源作为即付即用计算模型提供，并扩展到大型组织。就架构而言，该云可弹性扩展。例如大数据计算的 Google 云平台、Windows Azure HDInsight、云中的 Amazon 大数据计算及 RackSpace Cloudera Hadoop。

2）私有大数据云：这是组织内的云，旨在通过虚拟化基础架构提供隐私和更好的资源控制。

3）混合大数据云：该云拥护私有云和公共云的特征与功能。其目的是获得网络配置和延迟、可扩展性、高可用性、数据主权、灾难恢复、合规性和工作负载可移植性。当任务繁重时，其部署有助于将私有工作量迁移到公共体系结构。

4）大数据访问网络和计算平台：这是一个为数据、分析和计算而设计的集成平台。该平台由多个不同的提供商提供服务。

（3）元素和服务

大数据与云技术的融合产生了许多元素和服务。

1）云安装服务或安装云存储：这是一个旨在将数据安装到许多 Web 服务器和云存储的系统。例如亚马逊、谷歌驱动器。

2）大数据云：它是一种基础结构，旨在管理不同的数据资源（如数据管理、数据访问、安全性、调度程序、编程模型等）。大数据云基础架构具有许多工具，如流、Web 服务和 API，用于从不同来源（如 Google 数据、社交网络、关系数据存储等）收集数据。

3）云流传输：这是一种传输多媒体数据（如视频和音频）的方式，使多媒体数据可以在社交媒体基础结构上以连续模式轻松使用。

4）社交媒体：这些平台使在线收集大量信息成为可能。它是利用云技术服务的大数据的主要来源。例如 Facebook、LinkedIn、Twitter 和 YouTube。

5）HTTP、REST 服务：这些服务旨在为轻量级、可扩展和可维护的基于 Web 的应用程序开发 API。例如 USGS 地理云、谷歌地球。

6）大数据计算平台：它旨在创建管理不同数据源所需的模块，如数据密集型编程模型、数据安全、计算和数据感知调度、设备、分布式文件系统和分析。

7）计算云：开发该平台是为了提供所需的计算基础架构，如物理和虚拟计算以及来自私有、公共和混合云的存储。

## 2. 大数据云参考架构

大数据的云架构可有效管理复杂的计算可扩展性、存储和网络基础架构。作为服务提供商的基础架构除了存储应用程序外，还主要处理服务器、网络，并提供诸如虚拟化、基本监控和安全、操作系统、数据中心服务器以及存储服务之类的设施。以下讨论大数据云体系结构的四个层次。

（1）大数据分析——软件即服务（SaaS）

大数据分析以服务提供给用户，使他们具有了快速分析的能力，而无须投资基础结构和设施。该层的功能如下。

● 安排软件应用程序存储库。

● 在基础架构上部署软件程序。

● 结果交付给用户。

（2）大数据分析——平台即服务（PaaS）

这是体系结构的第二层。它是核心层，提供了平台相关服务以处理存储的大数据和进行计算。用于数据密集型和数据处理任务的数据管理工具、调度程序和编程环境（被视为中间件管理工具）都位于该区域中。该层负责提供开发分析所需的软件开发套件和工具。

（3）大数据结构（BDF）

这是大数据的结构层，负责寻址工具和 API，这些工具和 API 支持数据的存储、数据计算以及对不同应用程序服务的访问。该层包含 API 和可互操作的协议，旨在连接指定的多个云基础架构标准。

（4）云基础架构（CI）

云基础架构负责处理用于数据存储和计算即服务的基础架构。CI 层提供的服务如下。

- 创建用于大数据存储的大规模弹性基础结构，该基础结构能够按需部署。
- 设置动态虚拟机。
- 生成与基于文件、块和基于对象的大数据管理相关的按需存储功能。
- 使数据能够跨存储库无缝传递。
- 创建虚拟机并将文件系统安装到计算节点。

## 3. 云中大数据分析的好处

（1）改进的分析

随着云技术的进步，大数据分析变得更加完善，带来了更好的结果。因此，公司更喜欢在云中执行大数据分析。此外，云有助于整合来自众多来源的数据。

（2）简化的基础架构

大数据分析是基础架构上一项非常艰巨的工作，因为数据量大、速度快和类型各异，这是传统基础架构通常无法跟上的。由于云计算提供了灵活的基础架构，我们可以根据当时的需求进行扩展，因此很容易管理工作负载。

（3）降低成本

大数据和云技术都可以通过减少所有权为组织带来价值。云使客户无须大规模的大数据资源即可进行大数据处理。因此，大数据和云技术都在降低企业成本，并为企业带来价值。

（4）安全与隐私

数据安全和隐私是处理企业数据时的两个主要问题。一方面，当你的应用程序托管在云平台上时，由于其开放的环境和有限的用户控制，安全性成为首要问题。另一方面，作为一种开源应用程序，像 Hadoop 这样的大数据解决方案使用了大量的第三方服务和基础架构。因此，如今，系统集成商引入了具有弹性和可扩展性的私有云解决方案。此外，它还利用了可扩展的分布式处理。

除此之外，云数据在通常称为云存储服务器的中央位置存储和处理。服务提供商和客户将签署服务水平协议（SLA），以便彼此信任。如果需要，提供商还可以利用所需的高级安全控制级别。

与服务级别协议相关的规则可以保护数据、容量、可扩展性、安全、隐私以及数据存储的可用性和数据增长。

在许多组织中，大数据分析被用来检测和预防高级威胁与恶意黑客。

（5）虚拟化

基础架构在支持任何应用程序方面都起着至关重要的作用。虚拟化技术是大数据的理想平台。诸如 Hadoop 之类的虚拟化大数据应用程序具有多种优势，这些优势在物理基础架构上是无法得到的，但是它简化了大数据管理。大数据和云计算指出了各种技术与趋势的融合，这使 IT 基础架构和相关应用程序更加动态与模块化。因此，大数据和云计算项目非常依赖虚拟化。

# Text B

## What Is Cloud

A simple definition of cloud involves delivering different types of services over the Internet. From software and analytics to safe data storage and networking resources, everything can be delivered via the cloud.

### 1. Five characteristics of cloud

What exactly makes something a cloud? According to the National Institute of Standards and Technology (NIST), there are five defining characteristics of cloud.

(1) On-demand self-service

You can use it whenever you need it and pay per use. Think of it like electricity. In essence, the cloud is a form of utility computing. You create an account or pick your provider, and your services will be available to you anytime. You are billed at the end of the month only for what you used. This form of storing and accessing your data gives you full control over your resource usage and spending.

(2) Broad network access

You must be able to access from across the web using any device with internet connectivity. Wherever you are, your cloud data will be accessible through web browsers, on laptop or mobile devices. The reason for this is the fact that its underlying infrastructure includes servers on multiple locations.

(3) Resource pooling

Multiple tenants can share the same space and resources can be assigned, re-assigned and distributed as needed. You can be anywhere in the world and still have the equal access as everyone else, provided you have internet access.

(4) Rapid elasticity

Cloud can grow and shrink as much as possible without affecting any of its users or their information. For example, if your business is experiencing peak traffic, the cloud can expand to accommodate all the new requests.

(5) Measured service

You can examine how often people are using the cloud. Many cloud service providers utilize a pay-as-you-go model to ensure that their clients are getting what they pay for, no more and no less. Once again, this can be compared to electricity as you get billed for the amount that you use.

### 2. Types of cloud

There are three kinds of clouds, each with their unique benefits. You should evaluate cloud options to decide which is best for you and your business.

(1) Public cloud

Public cloud services are best for development systems and web servers. Your cloud computing provider will give you a slice of their digital space that they must share with other tenants.

This type of cloud is cost efficient since a pay-as-you-go model operates most. You pay for the number of hours you need to use the cloud and can exit whenever you complete your work. There are no obligations that require you to pay more than you need.

(2) Private cloud

Private clouds offer what their name suggests: privacy. You do not have to share your digital space with anyone else. Private cloud platforms are typically built in-house, and they belong to your business. They can also be configured in a third-party data center and still provide the advanced level of privacy.

Larger organizations and clients who are concerned about security favor private clouds. The reason for this is primarily the fact that private clouds offer more defense than their public counterparts. Companies who need to protect sensitive information like customer data rely on private clouds.

If you are using a private cloud, you know who has access to the data, you know if anyone made changes, and you know what to do in case of an emergency. You have full control over what happens to the cloud and don't have to worry about some third party vendor making changes that would negatively affect you. A firewall protects everything in your cloud from outsiders.

(3) Hybrid cloud

Hybrid clouds are the best of both worlds. If you are using a hybrid cloud, you can control an internal database and use the public cloud when needed. There might be times when you will need to move data and applications from the private cloud to the public cloud such as scheduled maintenance, blackouts and natural disasters. The ability to seamlessly migrate information is perfect for cloud disaster recovery solutions and preventing data loss.

The flexibility of hybrid clouds is excellent for scaling as any overflow can regulate in the public cloud. Furthermore, you can keep all non-sensitive tasks in the public cloud while safeguarding the essential data in the private cloud.

Regardless of how large your company is or what industry it serves, there will always be a cloud solution that fits your needs the best. Take the time to compare the advantages and disadvantages of each kind before deciding.

## 3. Benefits of cloud

(1) Always on available storage

The cloud provides an easy way to hold all your necessary data. You can rent cloud storage at a low price and scale it according to your demands. You no longer have to use an external hard drive or build an in-house data center.

(2) Disaster recovery solutions

You need data protection when catastrophe strikes. Preventing data loss as much as possible is critical regarding time, money and efficiency. Cloud provides a much faster and cost-effective disaster recovery than traditional solutions could ever offer.

Sometimes, the best way to deal with a disaster is to prepare for it beforehand. You should always consider worst-case scenarios since most catastrophic events are unplanned. Before cloud computing, you had to distribute and collect various tapes and drives and then transferred the data to a central location. Now, you can just click a few buttons and have it done for you.

(3) Cost saving

You no longer need to buy a ton of external hard drives to keep your critical information. Companies can save a lot of expenses annually by migrating data in the cloud. In addition to that, the cloud gives you access to advanced security systems and cutting-edge hardware and software, which adds up to the projected savings.

Cloud service providers that utilize a pay-as-you-go model are especially useful since you will never have to spend money on services that you are not using.

By contrast, with a monthly subscription service, you must pay to apply for the entire month regardless of how often you use it. If you use a monthly subscription service for only two weeks, you will get half of your money's worth.

(4) Consistent updates

The software is continuously being improved to increase security, efficiency, speed, capability and reliability. Software updates are consistent and usually don't need any extra costs.

(5) Business continuity

Ensuring business normal operation in case of a disaster is a significant challenge for most organizations. However, when a single minute of downtime can cost you more than implementing a backup and disaster recovery solution, business continuity management becomes a priority.

The cloud offers disaster recovery and business continuity solutions. You can rely on it to keep your data and applications active even if a disaster physically strikes your business. With a solid business continuity plan and right cloud solutions, you can minimize the effects of potential disruptions.

(6) Improved collaboration

You might work for a large international company with locations across the world. No matter where your offices are, every employee has the same access to relevant information via cloud technology. In addition, you can utilize cloud solutions by merely opening your phone.

Cloud collaboration tools offer important advantages to employees. They can make use of file versioning or real-time editing any time. They can access data, applications and services remotely from any device. All that boosts their productivity and, eventually, company's profits.

(7) Increased capacity

You no longer need to guess if you will have enough ability to build or destroy an application. Clouds can adjust upwards and downwards depending on what your business needs. The flexibility ensures that you will always be able to utilize cloud services regardless of what your business is doing.

(8) Performance and speed

The cloud commoditizes enterprise-grade technology, making it available to smaller companies as well. This form of utility computing makes emerging technologies available to businesses at an affordable price point.

You can access high-performance hardware and software to improve your operations. The opex-based delivery model makes cloud resources accessible to businesses of any sizes. You just need to pick the solution that meets your needs best.

(9) Data security

Keep your data secure and make sure that it does not fall into the wrong hands.

Cloud backups are an ideal solution to ensure business continuity and always-on availability of your files. All clouds offer some degree of encryption, deterrent and compliance, but private clouds remain the most secure from outsiders. Even so, you must beware of internal attacks.

**4. Types of cloud services**

Cloud services are as varied as the types of clouds themselves. You can purchase three different kinds of cloud services.

1) Infrastructure as a Service (IaaS) saves you money on buying physical data centers or servers. You pay as you go and only pay for as long as you need or use the service. IaaS allows you to adjust your scale depending on your demand quickly.

2) Platform as a Service (PaaS) has everything you need for your business applications. It comes complete with infrastructures such as networking, online storage and servers, as well as database management systems, development tools, etc. PaaS is designed to help create, test, develop and update your application.

3) Software as a Service (SaaS) is what you get whenever you download a new app for your phone. Companies create and develop their software and then lend it out to buyers. Businesses such as Autodesk, Lending Club, Microsoft and IBM all generate revenue from SaaS.

Figure out which service is best for you and your company. Cloud platforms are so diverse that it would be impossible to find a solution that didn't fit your needs.

## New Words

| | | |
|---|---|---|
| pick | [pɪk] | v.挑选 |
| provider | [prə'vaɪdə] | n.供应者，提供者 |
| pool | [pu:l] | n.池 |
| re-assigned | [rɪ ə'saɪnd] | adj.重新分配的，重新指定的 |
| accommodate | [ə'kɒmədeɪt] | vt.使适应 |
| | | vi.适应于；作调节 |
| option | ['ɒpʃn] | n.选择；选择权 |
| obligation | [,ɒblɪ'geɪʃn] | n.义务，责任；债务 |
| counterpart | ['kaʊntəpɑ:t] | n.副本；相对物；极相似的人或物 |
| negatively | ['negətɪvli] | adv.负面地，否定地，消极地 |
| firewall | ['faɪəwɔ:l] | n.防火墙 |
| blackout | ['blækaʊt] | n.停电 |
| migrate | [maɪ'greɪt] | vi.迁移，移往；移动 |
| overflow | [,əʊvə'fləʊ] | v.溢出 |
| regulate | ['regjuleɪt] | vt.调节，调整 |
| safeguarding | ['seɪfgɑ:dɪŋ] | n.安全措施 |
| catastrophe | [kə'tæstrəfi] | n.大灾难 |

| unplanned | [ˌʌn'plænd] | adj.无计划的，未筹划的 |
| priority | [praɪ'ɒrəti] | n.优先，优先权；重点 |
| potential | [pə'tenʃl] | adj.潜在的 |
| | | n.潜力；可能性 |
| destroy | [dɪ'strɔɪ] | vt.破坏，摧毁；使失败 |
| commoditize | [kə'mɒdɪtaɪz] | v.使商品化 |
| deterrent | [dɪ'terənt] | n.制止物；威慑物 |
| adjust | [ə'dʒʌst] | v.调整，调节 |

## ✍ Phrases

| on-demand self-service | 按需自助服务 |
| utility computing | 效用计算 |
| be available to | 可被……利用或得到 |
| pay-as-you-go model | 随用随付模式 |
| digital space | 数字空间 |
| cost efficient | 成本效益 |
| third-party data center | 第三方数据中心 |
| natural disaster | 自然灾害 |
| non-sensitive task | 非敏感任务 |
| worst-case scenario | 最坏的情况 |
| monthly subscription service | 每月订购服务 |
| opex-based delivery model | 基于运营的交付模型 |
| internal attack | 内部攻击 |
| online storage | 网络存储 |
| database management system | 数据库管理系统 |
| figure out | 计算出；弄明白；想出 |

## ✍ Abbreviations

| NIST(National Institute of Standards and Technology) | 国家标准技术研究院 |
| IaaS (Infrastructure as a Service) | 基础设施即服务 |
| SaaS (Software as a Service) | 软件即服务 |

## Reference Translation

## 什 么 是 云

云的简单定义指通过因特网交付不同类型的服务。从软件和分析到安全的数据存储与网络资源，一切都可以通过云交付。

**1. 云的五个特征**

云究竟是什么？根据美国国家标准技术研究院（NIST）的定义，云具有五个定义特征。

（1）按需自助服务

你可以在需要时使用它，并按使用量付费。把它想象成用电一样。本质上，云是效用计算的一种形式。你创建一个账户或选择你的提供商，即可随时享受服务。在月底你会收到账单要求为此付费。这种存储和访问数据的形式使你可以完全控制资源的使用与支出。

（2）广泛的网络访问

你必须能够使用任何具有因特网连接性的设备通过网络访问。无论你身在何处，都可以通过网络浏览器、笔记本计算机或移动设备访问云数据。这是因为其基础架构包括多个位置上的服务器。

（3）资源池

多个租户可以共享同一空间，可以根据需要分配、重新分配和分布资源。在世界上的任何地方，只要你可以访问互联网，就可以与其他人享有平等的访问权限。

（4）快速弹性

云可以在不影响其任何用户或其信息的情况下尽可能地增长和收缩。例如，如果你的企业遇到高峰流量，则云可以扩展以适应所有新请求。

（5）度量服务

你可以检查人们使用云的频率。许多云服务提供商采用随用随付模式来确保其客户按照所得付费，不多也不少。再次，当按照使用量计费时，可以比照电费。

## 2. 云的类型

云有三种类型，每种都有其独特的优势。你应该评估云选项，以确定最适合你和你的业务的选项。

（1）公共云

公共云服务最适合开发系统和网络服务器。云计算提供商将为你提供必须与其他租户共享的一个数字空间。

这种类型的云具有成本效益，因为主要以按需购买模型运行。你按照使用云的小时数付费，并且只要完成工作便可退出。无须支付超出所需的费用。

（2）私有云

私有云提供了其名称所暗示的含义：隐私。你不必与任何人共享数字空间。私有云平台通常由内部构建，它们属于你的企业。它们也可以在第三方数据中心中进行配置，并且仍然提供高级别的隐私。

关心安全性的较大型组织和客户倾向于私有云。其原因主要是私有云提供了比公共云更多的防御能力。需要保护敏感信息（如客户数据）的公司也依赖私有云。

如果你使用的是私有云，则可以知道谁有权访问数据、知道是否有人进行过更改，并且知道在紧急情况下该怎么办。你可以完全控制云发生的事情，而不必担心某些第三方供应商所做的更改会对你造成负面影响。防火墙可以保护你在云中的所有内容免受外部人员的攻击。

（3）混合云

混合云是两全其美的。如果使用混合云，则可以控制内部数据库并在需要时使用公共云。有时你可能需要将数据和应用程序从私有云移至公共云，如在计划维护、停电和发生自然灾害时。无缝迁移信息的能力非常适合云灾难恢复解决方案并防止数据丢失。

混合云的灵活性非常适合扩展，因为任何溢出都可以在公共云中进行调节。此外，你可以将所有非敏感任务保留在公共云中，而在私有云中保护重要数据。

无论你的公司规模有多大或服务的行业是什么，总会有一个最能满足你需求的云解决方案。在决定使用方案之前，建议花点时间比较其优点和缺点。

## 3. 云的好处

（1）存储始终可用

云提供了一种轻松的方法来保存所有必要的数据。你可以用低廉的价格租用云存储，然后根据需要进行扩展。你不再需要使用外部硬盘驱动器或构建内部数据中心。

（2）灾难恢复解决方案

灾难来临时，你需要保护数据。就时间、金钱和效率而言，尽可能防止数据丢失至关重要。与传统解决方案相比，云提供了更快、更具成本效益的灾难恢复。

有时，应对灾难的最好方法是事先做好准备。由于大多数灾难性事件都是计划外的，因此你应始终考虑最坏的情况。在以前没有用云计算的时候，你必须分发和收集各种磁带与驱动器，然后将数据传输到中央位置。现在，你只需单击几个按钮即可完成操作。

（3）节省成本

不再需要购买大量外部硬盘驱动器来保留你的重要信息。通过把数据迁移到云中，公司每年可以节省大量资金。除此之外，云使你可以使用先进的安全系统以及最先进的硬件和软件，从而可以节省预期成本。

利用随用随付模式的云服务提供商特别有用，因为你将永远不必在不使用的服务上花钱。

与之相比，每月订阅服务无论你使用频率如何，你都必须为整个月的申请付费。如果你仅使用两周的月度订阅服务，就浪费了一半的钱。

（4）持续更新

该软件不断得到改进，以提高安全性、效率、速度、功能和可靠性。软件更新是持续的，而且通常不需要任何额外费用。

（5）业务连续性

对于大多数组织而言，确保发生灾难时业务正常运行是一项重大挑战。但是，停机一分钟比实施备份和灾难恢复解决方案花费更大时，业务连续性管理就成为当务之急。

云提供了灾难恢复和业务连续性解决方案。即使灾难对企业造成了打击，你也可以依靠它来保持数据和应用程序处于活动状态。借助可靠的业务连续性计划和正确的云解决方案，你可以最大限度地减少潜在中断的影响。

（6）改善协作

你可能在一家全球都有分公司的大型国际公司工作。无论你的办公室在哪里，每个员工都可以通过云技术对相关信息进行相同的访问。此外，你只需打开手机即可使用云解决方案。

云协作工具为员工提供了重要优势。他们可以随时使用文件版本控制或实时编辑。他们可以从任何设备远程访问数据、应用程序和服务。所有这些都可以提高他们的生产力，并最终提高公司的利润。

（7）容量增加

你不再需要猜测是否有足够的能力来构建或终止应用程序。云可以根据你的业务需求上下调整。该灵活性确保无论你做什么业务，你都始终能够利用云服务。

（8）性能和速度

云将企业级技术商品化，让小型企业也可使用。这种形式的效用计算使新兴技术以可承受的价格提供给企业。

你可以访问高性能的硬件和软件以改善操作。基于运营的交付模型使各种规模的企业都可以访问云资源。只需要选择最能满足你需求的解决方案即可。

（9）数据安全

确保你的数据安全，并确保不会落入不当之人手中。

云备份是确保业务连续性和文件始终可用的理想解决方案。所有云都提供一定程度的加密、遏制力和合规性，但是私有云仍然对圈外人是最安全的。即使这样，你也必须提防内部攻击。

**4. 云服务的类型**

云服务的种类与云本身的种类一样多。你可以购买三种不同类型的云服务。

1）基础架构即服务（IaaS）为你节省了购买物理数据中心或服务器的费用。你随用随付，只在需要或使用该服务时才付费。IaaS 允许你根据需求快速调整规模。

2）平台即服务（PaaS）具有业务应用程序所需的一切。它完全配备了网络、在线存储和服务器等基础架构，以及数据库管理系统、开发工具等。PaaS 旨在帮助创建、测试、开发和更新你的应用程序。

3）每当你为手机下载新的应用程序时，即会获得软件即服务（SaaS）。公司创建和开发他们的软件，然后将其出租给买家。诸如 Autodesk、Lending Club、Microsoft 和 IBM 之类的企业均从 SaaS 产生收入。

找出最适合你和你公司的服务。云平台是如此之多，总能找到适合你需求的解决方案。

# Exercises

**[Ex. 1]** 根据 Text A 回答以下问题。

1. What is the purpose of big data clouds?

2. What are the features of big data clouds?

3. How many types are big data clouds classified into on the basis of their usage? What are they?

4. What is big data cloud?

5. What is big data computing platform designed to do?

6. What are the functions of the analytics of big data offered as service?

7. What is BDF responsible for?

8. What are the benefits of big data analysis in cloud?

9. What do the rules associated with service level agreements protect?

10. What do big data and cloud computing point to?

**[Ex. 2]** 根据 **Text B** 回答以下问题。

1. How many defining characteristics of cloud are there according to the National Institute of Standards and Technology (NIST)? What are they?

2. What is the reason for the fact that wherever you are, your cloud data will be accessible through web browsers, on laptop or mobile devices?

3. What do many cloud service providers utilize a pay-as-you-go model to ensure?

4. How many kinds of clouds are there?

5. Why do larger organizations and clients who are concerned about security favor private clouds?

6. What is the best way to deal with a disaster sometimes? Why should you always consider any worst-case scenarios?

7. What is a significant challenge for most organizations? When does business continuity management become a priority?

8. What can employees do with cloud collaboration tools?

9. What are the three different kinds of cloud services you can purchase?

10. What does Platform as a Service (PaaS) come complete with?

**[Ex. 3]** 词汇英译中
1. big data cloud         1. _____
2. utility computing         2. _____
3. computation as service     3. _____

4. pay-as-you-go model          4. _____

5. computing model              5. _____

6. deploy                       6. _____

7. disaster recovery            7. _____

8. duplication                  8. _____

9. distributed file system      9. _____

10. convergence                 10. _____

**[Ex. 4]** 词汇中译英

1. 存储库                       1. _____

2. 文件名                       2. _____

3. 大规模分布式计算             3. _____

4. 中间设备，中间件             4. _____

5. 物理基础设施                 5. _____

6. 多维                         6. _____

7. 虚拟机                       7. _____

8. 防火墙                       8. _____

9. 虚拟计算                     9. _____

10. 选择；选择权                10. _____

**[Ex. 5]** 短文翻译

**Features of Cloud Computing**

**1. Resources pooling**

It means that the cloud provider gathers the computing resources to provide services to multiple customers with the help of a multi-tenant model. There are different physical and virtual resources assigned and reassigned which depends on the demand of the customer. The customer generally has no control or information over the location of the provided resources, but he is able to specify location at a higher level of abstraction.

**2. On-demand self-service**

It is one of the important and valuable features of cloud computing as the user can continuously monitor the server uptime, capabilities and allotted network storage. With this feature, the user can also monitor the computing capabilities.

**3. Easy maintenance**

The servers are easily maintained and the downtime is very low and even in some cases, there is no downtime. Cloud computing comes up with an update every time by gradually making it better. The updates are more compatible with the devices and they perform faster than older ones.

## 4. Large network access

The user can access the data of the cloud or upload the data to the cloud from anywhere just with the help of a device and an internet connection. These capabilities are available all over the network and accessed with the help of internet.

## 5. Availability

The capabilities of the cloud can be modified as per the use and can be extended a lot. It analyzes the storage usage and allows the user to buy extra cloud storage if needed for a very small amount.

## 6. Automatic system

Cloud computing automatically analyzes the data needed and supports a metering capability at some level of services. We can monitor, control and report the usage. It will provide transparency for the host as well as the customer.

## 7. Security

Cloud security is one of the best features of cloud computing. It creates a snapshot of the data stored so that the data may not get lost even if one of the servers gets damaged. The data is stored within the storage devices, which cannot be hacked and utilized by any other person. The storage service is quick and reliable.

## 8. Pay as you go

In cloud computing, the user has to pay only for the service or the space they have utilized. There is no hidden or extra charge which is to be paid. The service is economical and most of the time some space is allotted for free.

# Reading Material

## Do You Want to Be a Big Data Analyst

With the increasing use of big data by organizations in every field, the need for big data analysts will continue to grow. Big data analysts examine vast amounts of varied data. They can uncover[1] hidden patterns, customer preferences[2] and market trends.

The goal of a big data analyst is to help organizations make better, more informed[3] decisions. Traditional data analysis cannot cope with the volume of big data, which includes both unstructured and structured data. Far more is needed than the ability to navigate relational databases and calculate statistical results. What big data analysts need most are the skills to translate relevant information into useful observations[4]. This requires technology to join hands with creativity[5], intuition[6] and

---

1  uncover [ʌnˈkʌvə] v.发现，揭示

2  preference [ˈprefrəns] n.喜好，偏爱

3  informed [ɪnˈfɔːmd] adj.明智的

4  observation [ˌɒbzəˈveɪʃn] n.观察；评论

5  creativity [ˌkriːeɪˈtɪvəti] n.创造性，创造力

6  intuition [ˌɪntjuˈɪʃn] n.直觉

experience.

The processing of big data is a relatively new field which can provide a competitive edge for businesses, and currently there is a shortage of big data analysts and data scientists.

Big data analytics applications are designed to analyze structured and unstructured data with the goal of identifying useful information, such as possible new sources of income or better marketing strategies. This process includes the analysis of internet click stream data, social media content, web server logs, text from customer emails and data captured by the Internet of Things[7].

## 1. Key skills needed by big data analysts

A big data analyst needs a broad range of skills to achieve their goals. Effective interpersonal skills[8] are quite useful in communicating big data results to employers and team members. Additionally, big data analysts should have the technical abilities needed for the work, which includes working with cloud services such as AWS (Amazon Web Services) or Microsoft Azure. It is not uncommon for a freelance big data analyst to have a preferred cloud service, but for security purposes, they may have to work with the company's cloud. Management skills can also be helpful in overseeing staff and working with assistants.

The following qualifications are generally expected from big data analysts.

- Industry experience: Analyzing big data requires an understanding of the industry, whether it be astronomy[9] or finance. This understanding provides a screening process or a paradigm, which is used to define and frame the questions being asked. The more experience one has in a particular field and life in general, the more understanding one will have when doing research. A broad background of experience provides an understanding of how to interpret[10] data.

- Statistics: Processing big data requires a knowledge of statistics. Statistics is a fundamental building block for data science, probability distribution[11] and random variables[12].

- Languages: For example Java, R, Python, C++, Hive, Ruby, SQL, MATLAB, SAS, SPSS, Weka, Scala and Julia. At a minimum, a big data analyst should be familiar with R, Python and Java.

- Computational[13] frameworks: Having a solid understanding of frameworks such as Apache Spark, Apache Samza, Apache Flink, Apache Storm and Hadoop is essential. These technologies support the processing of big data.

- Data warehousing: Understanding how data is stored and how to access it is important. Experience with non-relational database systems is also quite useful. Examples of non-relational (NoSQL) databases include Cassandra, HBase, CouchDB, HDFS and MongoDB.

---

7  Internet of Things: 物联网

8  interpersonal skill: 人际交往技巧

9  astronomy [əˈstrɒnəmi] n.天文学

10  interpret [ɪnˈtɜːprət] v.诠释；领会

11  probability distribution: 概率分布

12  random variable: 随机变量

13  computational [ˌkɒmpjuˈteɪʃənl] adj.计算的

- Data visualization: Big data can be difficult to comprehend and discuss. This is why pictures (also known as visuals) make discussing big data easier. Exploring even just a sample of data visualizing tools like Tableau or QlikView can show the shape of data, revealing hidden details.
- Communication: While data visualization is a useful tool, the ability to speak intelligently and clearly is a necessity for big data analysts. Results and how they were produced must be explained to the people paying the bill. After researching the data, big data analysts may also have to make presentations[14] to different departments within the organization.
- Written reports: A written report provides a permanent record of observations and conclusions for clients or employers.

## 2. Normal tasks and responsibilities

Big data analysts are responsible for realizing three key real-time solutions (affordability[15], speed and quality) and providing business intelligence to clients or employers. They may work with data quality teams ensuring data integrity and thoroughness[16], or perhaps with management to plan and perform data analyses. Big data analysts may also participate in planning organizational changes to maximize profits and minimize losses. Abhishek Mehta, the founder and CEO of Tresata, a predictive analytics company, stated, "The ability to deliver products and services at the right time, in the right place and to the right customer, instantly, is the future."

A big data analyst will regularly perform the following tasks.

- Determine organizational goals.
- Work with management, IT teams or data scientists.
- Mine data from a variety of sources.
- Screen and clean data to remove irrelevant information.
- Research trends and patterns.
- Find and identify new opportunities.
- Provide clear and concise data reports and visualizations for management.

---

14　presentation [ˌprezn'teɪʃn] n.演示，展示; 介绍
15　affordability [əˌfɔːdə'bɪləti] n.支付能力; 承受能力
16　thoroughness ['θʌrənəs] n.完整性

# Unit 10

## Text A

扫码听课文

### Big Data Security

The adoption of big data analytics is rapidly growing. Sensitivities around big data security and privacy are a hurdle that organizations need to overcome.

### 1. What is big data security

Big data security is the collective term for all the measures and tools used to guard both the data and analytics processes from attacks, theft, or other malicious activities that could harm or negatively affect them. Much like other forms of cyber security, the big data security is concerned with attacks that originate either from the online or offline spheres.

Big data security is the processing of guarding data and analytics processes, both in the cloud and on-premise, from any number of factors that could compromise their confidentiality.

For companies that operate on the cloud, big data security challenges are multi-faceted.

These challenging threats include the theft of information stored online, ransomware, XSS (Cross Site Scripting) attacks or DDoS (Distributed Denial of Service) attacks that could crash a server. Attacks on an organization's big data storage could cause serious financial repercussions such as losses, litigation costs and fines or sanctions.

### 2. Key big data security issues

The following is a shortlist of some of the obvious big data security issues (or available tech) that should be considered.

- Distributed frameworks. Most big data implementations actually distribute huge processing jobs across many systems for faster analysis. Distributed processing may mean less data processed by any one system, but it means a lot more systems where security issues can crop up.
- Non-relational data stores. Non-relational databases like NoSQL databases usually lack security by themselves.
- Storage. In big data architecture, the data is usually stored on multiple tiers, depending on business needs for performance and cost. For instance, high-priority "hot" data will usually be stored on flash media. So locking down storage will mean creating a tier-conscious strategy.
- Endpoints. Security solutions that draw logs from endpoints will need to validate the authenticity of those endpoints, or the analysis isn't going to do much good.
- Real-time security/compliance tools. These tools generate a tremendous amount of information. The key is to find a way to ignore the false or rough information, so human

talent can be focused on the true breaches or valuable information.

- Data mining solutions. These solutions are the heart of many big data environments. They find the patterns that suggest business strategies. For that very reason, it's particularly important to ensure they're secured against not just external threats, but insiders who abuse network privileges to obtain sensitive information.

- Access controls. It's critically important to provide an encrypted authentication/validation system to verify that users are who they say they are and determine who can see what.

- Granular auditing can help determine when missed attacks occurred, what the consequences were and what should be done to improve matters in the future. This in itself is a lot of data, and must be enabled and protected to be useful in addressing big data security issues.

- Data provenance primarily concerns metadata (data about data), which can be extremely helpful in determining where data came from, who accessed it, or what was done with it. Usually, this kind of data should be analyzed with exceptional speed to minimize the time in which a breach is active. Privileged users engaged in this type of activity must be thoroughly vetted and closely monitored to ensure they don't become their own big data security issues.

## 3. Big data security challenges

There are several challenges to securing big data that can compromise its security. Keep in mind that these challenges are by no means limited to on-premise big data platforms. They also pertain to the cloud. When you host your big data platform in the cloud, take nothing for granted. Work closely with your provider to overcome these same challenges with strong security service level agreements.

Typical challenges to securing big data.

- Advanced analytic tools for unstructured big data and non-relational databases (NoSQL) are newer technologies in active development. It can be difficult for security software and processes to protect these new toolsets.

- Mature security tools effectively protect data ingress and storage. However, they may not have the same impact on data output from multiple analytics tools to multiple locations.

- Big data administrators may decide to mine data without permission or notification. Whether the motivation is curiosity or criminal profit, your security tools need to monitor and alert on suspicious access no matter where it comes from.

- The sheer size of a big data installation, terabytes to petabytes large, is too big for routine security audits. And because most big data platforms are cluster-based, this introduces multiple vulnerabilities across multiple nodes and servers.

- If the big data owner does not regularly update security for the environment, they are at risk of data loss and exposure.

- Security tools need to monitor and alert on suspicious malware infection on the system, database or a web CMS (Content Management System) such as WordPress, and big data security experts must be proficient in cleanup and know how to remove malware.

## 4. Big data security technologies

None of these big data security tools are new. What is new is their scalability and the ability to secure multiple types of data in different stages.

- Encryption: Your encryption tools need to secure data in transit and at rest, and they need to do it across massive data volumes. Encryption also needs to operate on many different types of data, both user-generated and machine-generated. Encryption tools also need to work with different analytics toolsets and their output data. These data are storage in Relational DataBase Management Systems (RDBMS), non-relational databases like NoSQL, and specialized file systems such as Hadoop Distributed File System (HDFS) in common big data storage formats.

- Centralized key management: Centralized key management has been a best security practice for many years. It is applied in big data environments, especially in those with wide geographical distribution. Best practices include policy-driven automation, logging, on-demand key delivery, and abstracting key management from key usage.

- User access control: User access control may be the most basic network security tool, but many companies practice minimal control because the management overhead can be so high. Strong user access control requires a policy-based approach that automates access based on user and role-based settings. Policy-driven automation manages complex user control levels, such as multiple administrator settings that protect the big data platform against inside attack.

- Intrusion detection and prevention: Intrusion detection and prevention systems are security workhorses. This does not make them any less valuable to the big data platform. Big data's value and distributed architecture lend themselves to intrusion attempts. IPS (Intrusion Prevention System) enables security admins to protect the big data platform from intrusion, and should an intrusion succeed, IDS (Intrusion Detection System) quarantine the intrusion before it does significant damage.

- Physical security: Don't ignore physical security. Build it in when you deploy your big data platform in your own data center, or carefully do due diligence around your cloud provider's data center security. Physical security systems can deny data center access to strangers or to staff members who have no business being in sensitive areas. Video surveillance and security logs will do the same.

## 5. How can you implement big data security

Organizations can protect their big data analytics tools with a variety of security measures. One of the most common security tools is encryption, a relatively simple tool that can go a long way. Encrypted data is useless to external actors such as hackers if they don't have the key to unlock it. Moreover, encrypting data means that both at input and output, information is completely protected.

Building a strong firewall is another useful big data security tool. Firewalls are effective at filtering traffic that both enters and leaves servers. Organizations can prevent attacks before they happen by creating strong filters that avoid any third parties or unknown data sources.

Data security must complement other security measures such as endpoint security, network security, application security, physical site security and more to create an in-depth approach. By planning ahead and being prepared for the introduction of big data analytics in your organization, you will be able to help your organization meet its objectives securely.

## ✎ New Words

| | | |
|---|---|---|
| adoption | [ə'dɒpʃn] | n.采用 |
| sensitivity | [ˌsensə'tɪvəti] | n.敏感；感受性 |
| privacy | ['prɪvəsi] | n.隐私，秘密 |
| overcome | [ˌəʊvə'kʌm] | v.克服；战胜 |
| guard | [gɑ:d] | v.保护；守卫 |
| attack | [ə'tæk] | v.&n.攻击，侵袭；损害 |
| malicious | [mə'lɪʃəs] | adj.恶意的；蓄意的；预谋的 |
| harm | [hɑ:m] | v.&n.伤害，危害 |
| cyber | ['saɪbə] | adj.计算机（网络）的 |
| variant | ['veəriənt] | n.变体，变种，变形 |
| sphere | [sfɪə] | n.范围 |
| on-premise | [ɒn'premɪs] | adj.本地的；预置的 |
| compromise | ['kɒmprəmaɪz] | v.损害；违背；使陷入危险 |
| | | n.折中，妥协 |
| confidentiality | [ˌkɒnfɪˌdenʃi'æləti] | n.机密性 |
| multi-faceted | ['mʌlti 'fæsitid] | adj.多方面的、多元化的 |
| ransomware | ['rænsəmweə] | n.勒索软件 |
| crash | [kræʃ] | n.崩溃；碰撞 |
| | | v.（使）崩溃，（使）瘫痪 |
| repercussion | [ˌri:pə'kʌʃn] | n.后果；反响 |
| fine | [faɪn] | n.罚款 |
| | | v.对……处以罚金 |
| sanction | ['sæŋkʃn] | n.制裁，处罚 |
| obvious | ['ɒbviəs] | adj.明显的；当然的 |
| tech | [tek] | n.技术 |
| multiple | ['mʌltɪpl] | adj.多重的；多个的；多功能的 |
| flash | [flæʃ] | n.闪存 |
| validate | ['vælɪdeɪt] | vt.确认；证实 |
| authenticity | [ˌɔ:θen'tɪsəti] | n.真实性；可靠性 |
| ignore | [ɪg'nɔ:] | v.忽略，不理 |
| rough | [rʌf] | adj.粗略的 |
| suggest | [sə'dʒest] | v.建议；推荐；表明 |

| abuse | [ə'bju:s] | n.&v.滥用 |
| privilege | ['prɪvəlɪdʒ] | n.特权；优惠 |
| critically | ['krɪtɪkli] | adv.危急地；严重地 |
| authentication | [ɔ:ˌθentɪ'keɪʃn] | n.身份验证；认证 |
| consequence | ['kɒnsɪkwəns] | n.结果；重要性 |
| provenance | ['prɒvənəns] | n.起源，出处 |
| exceptional | [ɪk'sepʃənl] | adj.杰出的，非凡的 |
| privileged | ['prɪvəlɪdʒd] | adj.享有特权的；特许的，专用的 |
| toolset | ['tu:lset] | n.工具集，工具包，工具箱；成套工具 |
| mature | [mə'tʃʊə] | adj.成熟的；到期的 |
| ingress | ['ɪngres] | n.进入，进入权 |
| administrator | [əd'mɪnɪstreɪtə] | n.管理者 |
| permission | [pə'mɪʃn] | n.准许；许可证 |
| motivation | [ˌməʊtɪ'veɪʃn] | n.动力；积极性 |
| curiosity | [ˌkjʊəri'ɒsəti] | n.好奇，求知欲 |
| suspicious | [sə'spɪʃəs] | adj.可疑的；怀疑的；不信任的 |
| terabyte | ['terəbaɪt] | n.太字节，缩写为 TB，1TB=1024GB=$2^{40}$B |
| petabyte | ['petəbaɪt] | n.拍字节，缩写为 PB，1PB=1024TB=$2^{50}$B |
| audit | ['ɔ:dɪt] | n.&v.审计；检查 |
| cluster-based | ['klʌstə beɪst] | adj.基于集群的 |
| introduce | [ˌɪntrə'dju:s] | vt.引进；提出；介绍 |
| vulnerability | [ˌvʌlnərə'bɪləti] | n.弱点；脆弱性 |
| update | [ˌʌp'deɪt] | vt.更新，升级 |
| | [ʌp'deɪt] | n.更新，升级 |
| exposure | [ɪk'spəʊʒə] | n.暴露 |
| malware | ['mælweə] | n.恶意软件，流氓软件 |
| proficient | [prə'fɪʃnt] | adj.精通的，熟练的 |
| scalability | [skeɪlə'bɪləti] | n.可扩展性；可升级性；可伸缩性；可量测性 |
| encryption | [ɪn'krɪpʃn] | n.编密码；加密 |
| geographical | [ˌdʒi:ə'græfɪkl] | adj.地理的 |
| delivery | [dɪ'lɪvəri] | n.交付 |
| overhead | ['əʊvəhed] | n.经常性支出，运营费用 |
| policy-based | ['pɒləsi beɪst] | adj.基于策略的 |
| role-based | [rəʊl beɪst] | adj.基于角色的 |
| protect | [prə'tekt] | v.保护 |
| intrusion | [ɪn'tru:ʒn] | n.入侵 |

| | | |
|---|---|---|
| detection | [dɪ'tekʃn] | n.检测，探测 |
| prevention | [prɪ'venʃn] | n.预防；阻止 |
| quarantine | ['kwɒrənti:n] | vt.对……进行检疫；隔离 |
| damage | ['dæmɪdʒ] | v.&n.损坏，伤害 |
| deny | [dɪ'naɪ] | v.否认；拒绝 |
| surveillance | [sɜ:'veɪləns] | n.监视；监督 |
| traffic | ['træfɪk] | n.信息流；流量，通信量 |
| filter | ['fɪltə] | n.过滤器 |

## ✍ Phrases

| | |
|---|---|
| big data security | 大数据安全 |
| collective term | 统称 |
| negatively affect | 负面影响，消极影响 |
| be concerned with | 涉及；参与，干预；关心 |
| originate from | 来自……，源于…… |
| litigation cost | 诉讼费 |
| crop up | 犯错误；突然发生；意外地发现 |
| tier-conscious strategy | 层级意识策略 |
| external threat | 外部威胁 |
| sensitive information | 敏感信息 |
| access control | 访问控制 |
| be engaged in | 正着手于，正做着，致力于 |
| limit to | （把……）限制在……，局限于…… |
| pertain to | 属于；关于；适合 |
| take ... for granted | 认为……理所当然，想当然；轻信 |
| security service level agreement | 安全服务级别协议 |
| in transit | 传输中的；在途中 |
| centralized key management | 集中式密钥管理 |
| user access control | 用户访问控制 |
| due diligence | 应有的注意，尽职调查 |

## ✍ Abbreviations

| | |
|---|---|
| XSS (Cross Site Scripting) | 跨站脚本，为了与 CSS（Cascading Style Sheets）区别，改了缩写 |
| DDoS (Distributed Denial of Service) | 分布式拒绝服务 |
| CMS (Content Management System) | 内容管理系统 |

IPS (Intrusion Prevention System)      入侵预防系统

IDS (Intrusion Detection System)      入侵检测系统

# Reference Translation

## 大数据安全

大数据分析的应用正在迅速增长。围绕大数据安全和隐私的敏感性是组织需要克服的障碍。

### 1. 什么是大数据安全

大数据安全是所有用于保护数据和分析过程免受可能损害或负面影响的攻击、盗窃或其他恶意活动的措施与工具的统称。就像其他形式的网络安全一样，大数据安全关注的是源自在线或离线领域的攻击。

大数据安全是对云上和本地的数据与分析过程的保护，这可使得任何因素都不能损害其机密性。

对于在云上运行的公司，大数据安全的挑战是多方面的。

这些具有挑战性的威胁包括盗窃在线存储的信息、运行勒索软件、XSS（跨站脚本）攻击或能使服务器崩溃的 DDoS（分布式拒绝服务）攻击。对组织中大数据存储的攻击可能会导致严重的财务后果，如损失资金和诉讼费用以及罚款或制裁。

### 2. 大数据安全的关键问题

以下是应考虑的一些明显的大数据安全问题（或可用技术）的清单。

- 分布式框架。实际上，大多数的大数据处理操作都会把大的处理任务分配给许多系统，以加快分析速度。分布式处理可能意味着任何一个系统处理的数据更少，但这意味着可能出现安全问题的系统的数量更多。
- 非关系型数据存储。像 NoSQL 数据库这样的非关系型数据库通常本身就缺乏安全性。
- 存储。在大数据架构中，数据通常存储在多个层中，这取决于业务对性能和成本的需求。例如，高优先级的"热"数据通常会存储在闪存介质上。因此，锁定存储将意味着创建层级意识策略。
- 端点。从端点提取日志的安全解决方案将需要验证那些端点的真实性，否则分析将用处不大。
- 实时安全/合规工具。这些工具会产生大量信息。关键是找到一种忽略虚假或粗略信息的方法，从而人们可以专注于真正的漏洞或有价值的信息。
- 数据挖掘解决方案。这些解决方案是许多大数据环境的核心。它们找到建议业务策略的模式。因此，不仅要保护它们不受外部威胁，而且要防止内部人员滥用网络特权来获取敏感信息，这一点尤为重要。
- 访问控制。提供一个加密的身份认证/验证系统来验证用户的身份并确定谁可以看到什么，这一点至关重要。
- 粒度审核可以帮助你确定已经错过的攻击出现的时间、分析后果以及将来应采取哪些措施来改善。这本身就有许多数据，必须激活和保护这些数据才能在解决大数据安全问题时有用。

- 数据来源主要涉及元数据（有关数据的数据），这对于确定数据的来源、访问者或对数据的处理非常有帮助。通常，应以极快的速度分析此类数据，以最大程度缩短漏洞的活跃时间。必须仔细审查并严格监控从事此类活动的特权用户，以确保他们不会成为自己的大数据安全问题。

## 3. 大数据安全挑战

保护大数据不受危害可能面临许多安全挑战。请记住，这些挑战绝不仅限于本地大数据平台。它们也与云有关。当你将大数据平台托管在云中时，不应心存侥幸。可以通过强大的安全服务级别协议，与你的提供商紧密合作来克服这些挑战。

保护大数据的典型挑战如下。

- 用于非结构化大数据和非关系数据库（NoSQL）的高级分析工具是一个较新的技术，正处于积极开发阶段。安全软件和过程可能很难保护这些新工具集。
- 成熟的安全工具可有效保护数据进入和存储。但是，它们可能对从多个分析工具到多个位置的数据输出产生不同的影响。
- 大数据管理员可以决定在未经许可或通知的情况下挖掘数据。无论是出于好奇还是出于犯罪目的，你的安全工具都需要监控可疑访问并发出警报，无论它来自何处。
- 仅安装大数据就需要 TB 到 PB 级的存储空间，这对常规的安全审核太大了。而且由于大多数的大数据平台都是基于集群的，因此会在多个节点和服务器上引入多个漏洞。
- 如果大数据所有者没有定期更新环境安全，则存在数据丢失和暴露的风险。
- 安全工具需要监控系统、数据库或像 WordPress 这样的 Web 内容管理系统，当出现可疑恶意软件感染时发出警报，大数据安全专家必须精通清理工作，并且知道如何删除恶意软件。

## 4. 大数据安全技术

这些大数据安全工具都不是新的。它们的可扩展性以及在不同阶段保护多种类型数据的能力是新增的。

- 加密：你的加密工具需要保护传输中的数据和静态的数据，并且需要跨海量数据进行处理。还需要对用户和计算机生成的许多不同类型的数据进行加密。加密工具还需要与不同的分析工具集及其输出数据一起使用，这些数据以常用的大数据格式存储于关系数据库管理系统（RDBMS）、非关系数据库（如 NoSQL）和专用文件系统（如 Hadoop 分布式文件系统（HDFS））中。
- 集中式密钥管理：多年来，集中式密钥管理一直是最佳的安全实践。它适用于大数据环境，尤其是地理分布广泛的环境。最佳实践包括策略驱动的自动化、日志记录、按需密钥交付以及从密钥用法中提取密钥管理。
- 用户访问控制：用户访问控制可能是最基本的网络安全工具，但是许多公司实行的控制很少，因为管理开销可能会很高。强大的用户访问控制需要基于策略的方法，该方法可基于用户和基于角色的设置自动进行访问。策略驱动的自动化管理复杂的用户控制级别，如多个管理员设置可保护大数据平台免受内部攻击。
- 入侵检测和防御：入侵检测和防御系统是安全的主力军。这不会使它们降低大数据平台的价值。大数据的价值和分布式体系结构对入侵尝试有利。IPS（入侵防御系统）使安全管理员可以保护大数据平台免受入侵，如果入侵成功，IDS（入侵检测系统）

会在入侵造成重大损害之前对其进行隔离。

- 物理安全：不要忽略物理安全。当你在自己的数据中心中部署大数据平台时要构建物流安全，或对云提供商的数据中心安全进行认真调查。物理安全系统可以拒绝陌生人或在敏感区域没有业务的员工访问数据中心。视频监控和安全日志将执行相同的操作。

## 5. 如何实施大数据安全

组织可以通过多种安全措施来保护其大数据分析工具。加密是最常见的安全工具之一，它是一种相对简单的工具，可以使用很长时间。如果外部参与者（如黑客）没有解锁密钥，则对其而言加密数据毫无用处。此外，对数据进行加密意味着输入和输出的信息均受到完全保护。

建立强大的防火墙是另一个有用的大数据安全工具。防火墙可以有效过滤进入和离开服务器的流量。组织可以通过创建可避免任何第三方或未知数据源的强大过滤器来防止攻击发生。

数据安全必须与其他安全措施（如端点安全、网络安全、应用程序安全、物理站点安全等）相辅相成，以创建一种深入的方法。通过提前计划并为在组织中引入大数据分析做好准备，你将能够帮助组织安全地实现其目标。

# Text B

扫码听课文

## Big Data Privacy

Big data privacy involves properly managing big data to minimize risk and protect sensitive data. Because big data comprises large and complex data sets, many traditional privacy processes cannot handle the scale and velocity required. To safeguard big data and ensure it can be used for analytics, you need to create a framework for privacy protection that can handle the volume, velocity, variety and value of big data as it is moved between environments, processed, analyzed and shared.

## 1. Big data privacy

In an era of hybrid, multi-cloud computing, data owners must keep up with both the pace of data growth and the proliferation of regulations that govern it — especially regulations protecting the privacy of sensitive data and Personally Identifiable Information (PII). With more data spreading across more locations, the business risk of a privacy breach has never been higher, with consequences ranging from high fines to loss of market share.

Big data privacy is also a matter of customer trust. The more data you collect about users, the easier it gets to "connect the dots" to understand their current behavior, draw inferences about their future behavior, and eventually develop deep and detailed profiles of their lives and preferences. The more data you collect, the more important it is to be transparent with your customers about what you're doing with their data, how you're storing it, and what steps you're taking to comply with regulations that govern privacy and data protection.

The volume and velocity of data from existing sources is expanding fast. You also have new (and growing) varieties of data types and sources, such as social networks and IoT device streams. To keep pace, your big data privacy strategy needs to expand, too. That requires you to consider all of these issues.

- What do you intend to do with customer and user data?
- How accurate is the data, and what are the potential consequences of inaccuracies?
- How will your data security scale to keep up with threats of data breaches and insider threats as they become more common?
- Where is your balancing point between the need to keep data locked down in-place and the need to expose it safely so you can extract value from it?
- How do you maintain compliance with data privacy regulations that vary across the countries and regions where you do business, and how does that change based on the type or origin of the data?
- How do you maintain transparency about what you do with the big data you collect without giving away the "secret sauce" of the analytics that drive your competitive advantage?

## 2. Predictions for big data privacy

Prediction 1: Data privacy mandates will become more common.

As organizations store more types of sensitive data in larger amounts over longer periods of time, they will be under increasing pressure to be transparent about what data they collect, how they analyze and use it, and why they need to retain it. The European Union's General Data Privacy Regulation (EUGDPR) is a high-profile example. More government agencies and regulatory organizations are following suit. To respond to these growing demands, companies need reliable and scalable big data privacy tools that encourage and help people to access, review, correct, anonymize and even purge some or all of their personal and sensitive information.

Prediction 2: New big data analytic tools will enable organizations to perform deeper analysis of legacy data, discover uses for which the data wasn't originally intended, and combine it with new data sources.

Big data analytics tools and solutions can now dig into data sources that were previously unavailable, and identify new relationships hidden in legacy data. That's a great advantage when it comes to getting a complete view of your enterprise data.

The key to protecting the privacy of your big data while still optimizing its value is the ongoing review of the four critical data management activities.

- Data collection.
- Retention and archiving.
- Data use.
- Creating and updating disclosure policies and practices.

Companies with a strong and scalable data governance program will have an advantage when assessing these tasks. They will be able to accurately assess data-related risks and benefits in less time and quickly take more decisive action based on trusted data.

## 3. Big data privacy and protection strategies

Traditional data security is network and system-centric, but today's hybrid and multi-cloud architectures spread data across more platform agnostic locations and incorporate more data types than ever before. Big data privacy can't be an afterthought. It must be an integral part of your cloud integration and data management strategy.

- You must define and manage data governance policies to clarify what data is critical and why, who owns the critical data, and how it can be used responsibly.

- You must discover, classify and understand a wide range of sensitive data across all big data platforms at massive scale by leveraging artificial intelligence and machine learning tools to automate controls. Then, you can use that information to develop and implement intelligent big data management policies.

- You must index, inventory and link data subjects and identities to support data access rights and notifications.

- You must be able to perform continuous risk analysis for sensitive data to understand your risk exposure, prioritize available data protection resources and investments, develop protection and remediation plans as your big data grows.

- You need automated, centralized big data privacy tools that integrate with native big data tools like Cloudera Sentry, Amazon Macie and Hortonworks Ranger to streamline and facilitate the process of managing data access, such as viewing, changing and adding access policies.

- You need fast and efficient data protection capabilities at scale, including dynamic masking for big data as it's put into use in production and data lakes, encryption for big data at rest in data lakes and data warehouses and persistent masking for big data used in non-production environments.

- You must measure and communicate the status of big data privacy risk indicators, which is a critical part of tracking success in protecting sensitive information while supporting audit readiness.

## New Words

| | | |
|---|---|---|
| minimize | ['mɪnɪmaɪz] | vt.把……减至最低数量（程度） |
| safeguard | ['seɪfgɑːd] | n.保护；防护措施；安全设施 |
| | | vt.保护；防护 |
| hybrid | ['haɪbrɪd] | n.混合物 |
| multi-cloud | ['mʌltɪ klaʊd] | n.多云 |
| proliferation | [prəˌlɪfə'reɪʃn] | n.增长 |
| inference | ['ɪnfərəns] | n.推理，推断，推论 |
| preference | ['prefrəns] | n.偏爱；优先权 |
| transparent | [træns'pærənt] | adj.透明的；易懂的 |

| | | |
|---|---|---|
| inaccuracy | [ɪn'ækjərəsi] | n.不准确，误差 |
| mandate | ['mændeɪt] | n.授权；委任 |
| period | ['pɪəriəd] | n.一段时间；阶段；时期 |
| pressure | ['preʃə] | n.压力 |
| agency | ['eɪdʒənsi] | n.服务/代理/经销机构；（政府）专门机构 |
| encourage | [ɪn'kʌrɪdʒ] | v.鼓励；促进 |
| anonymize | [ə'nɒnɪmaɪz] | vt.使匿名 |
| purge | [pɜ:dʒ] | vt.清除，（使）净化<br>n.净化 |
| unavailable | [ˌʌnə'veɪləbl] | adj.难以获得的；不能利用的 |
| retention | [rɪ'tenʃn] | n.保留；保持力 |
| governance | ['gʌvənəns] | n.治理，管理 |
| afterthought | ['ɑ:ftəθɔ:t] | n.事后的考虑或想法 |
| responsibly | [rɪ'spɒnsəbli] | adv.负责地；有责任感地 |
| subject | ['sʌbdʒɪkt] | n.主题<br>v.提出 |
| identity | [aɪ'dentəti] | n.身份；个性 |
| prioritize | [praɪ'ɒrətaɪz] | vt.按重要性排列，划分优先顺序；优先处理 |
| investment | [ɪn'vestmənt] | n.投资；（时间/精力的）投入 |
| remediation | [rɪˌmi:dɪ'eɪʃn] | n.补救；纠正 |
| automated | ['ɔ:təmeɪtɪd] | adj.自动化的 |
| centralized | ['sentrəlaɪzd] | adj.集中的 |
| native | ['neɪtɪv] | adj.本地的 |
| streamline | ['stri:mlaɪn] | vt.使简单化；使现代化 |
| persistent | [pə'sɪstənt] | adj.持续的；坚持不懈的；持久的 |
| readiness | ['redɪnəs] | n.准备就绪；愿意，乐意 |

## ✍ Phrases

| | |
|---|---|
| market share | 市场份额 |
| connect the dots | 连点成线 |
| social network | 社交网络 |
| lock down | 锁定，保持 |
| privacy regulation | 隐私法规 |
| secret sauce | 秘诀 |
| follow suit | 跟着做，照着做；如法炮制 |
| respond to | 对……做出反应 |
| platform agnostic | 平台无关的 |

artificial intelligence          人工智能
risk exposure          风险敞口
dynamic masking          动态屏蔽
at rest          静止，不动

## ✍ Abbreviations

PII (Personally Identifiable Information)          个人可识别的信息
IoT (Internet of Things)          物联网
EUGDPR (European Union's General Data Privacy Regulation)          欧盟的通用数据隐私法规

## Reference Translation

## 大数据隐私

大数据隐私涉及正确管理大数据，以便最大限度地降低风险并保护敏感数据。由于大数据包含庞大而复杂的数据集，因此许多传统的隐私保护流程无法处理所需的规模和速度。为了保护大数据并确保其可用于分析，你需要创建一个隐私保护框架，当数据在不同环境中移动，对大数据进行处理、分析和共享时，该框架可以保护大数据的数量、速度、多样性和价值。

### 1. 大数据隐私

在混合多云计算的时代，数据所有者必须跟上数据增长和管理数据法规的增长，特别是适应保护敏感数据和个人身份信息（PII）隐私的法规。随着越来越多的数据分布在更多的位置，侵犯隐私的商业风险从未如此之高，其后果包括从受到高额罚款直至失去市场份额。

大数据隐私也是客户信任的问题。你收集的有关用户的数据越多，就越容易"连点成线"了解他们的当前行为、推断他们的未来行为，并最终深入了解他们的生活和偏好。你收集的数据越多，与你的客户保持透明的关系就越重要，那就是让客户了解你对数据的处理方式、存储方式以及为遵守有关隐私和数据的法规所采用的步骤。

来自现有资源的数据量和速度正在迅速增长。你还拥有新的（并且正在增长的）各种数据类型和数据源，如社交网络和物联网设备流。为了跟上步伐，你的大数据隐私策略也需要扩展。这就需要考虑以下这些问题。

- 你打算如何处理客户和用户数据？
- 数据的准确性如何？数据不准确的潜在后果是什么？
- 随着数据泄露威胁和内部威胁变得越来越普遍，你的数据安全将如何扩展？
- 如何平衡在适当位置保持数据与为获取价值而安全地公开数据？
- 如何遵守业务开展所在国家和地区的数据隐私法规，以及如何根据数据的类型或来源进行更改？
- 你如何在不泄露提高你竞争优势的分析"秘诀"的情况下，保持你对所收集的大数据的处理透明度？

### 2. 大数据隐私的预测

预测 1：数据隐私授权将变得更加普遍。

随着组织在更长的时间段内大量存储更多类型的敏感数据，它们将承受越来越大的压力，要求透明化收集的数据、说明如何分析和使用数据，以及为什么需要保留数据。欧盟的通用数据隐私法规（EUGDPR）是一个引人注目的例子。越来越多的政府机构和监管组织也纷纷效仿。为了满足这些不断增长的需求，公司需要可靠、可扩展的大数据隐私工具，以鼓励和帮助人们访问、查看、更正、匿名化甚至清除部分或全部的个人和敏感信息。

预测 2：新的大数据分析工具将使组织能够对遗留数据进行更深入的分析，发现数据原来未期望的用途，并将其与新的数据源结合起来。

大数据分析工具和解决方案现在可以挖掘以前不可用的数据源，并识别隐藏在旧数据中的新关系。当谈论到完整的企业数据视图时，这是一个很大的优势。

在保护大数据隐私的同时仍可优化其价值的关键是不断审查以下四个关键数据管理活动。

- 数据采集。
- 保留和存档。
- 数据使用。
- 创建和更新披露政策与实践。

拥有强大、可扩展的数据治理计划的公司在评估这些任务时将具有优势。它们将能够在更短的时间内准确评估与数据相关的风险和收益，并根据可信赖的数据迅速采取更具决定性的行动。

### 3. 大数据隐私与保护策略

传统的数据安全性是以网络和系统为中心的，但是如今的混合和多云体系结构将数据分布在与平台无关的更多位置，并且合并的数据类型比以往任何时候都要多。大数据隐私不是事后的想法。它必须是云集成和数据管理策略的组成部分。

- 你必须定义和管理数据治理策略，以阐明哪些数据至关重要、为什么、谁拥有该关键数据以及如何负责任地使用它们。
- 你必须利用人工智能和机器学习工具来自动化控制，从而发现、分类和理解所有大数据平台上的各种敏感数据。然后，可以使用该信息来制定和实施智能大数据管理策略。
- 你必须索引、清查并链接数据主题和身份，以支持数据访问权限和通知。
- 你必须能够对敏感数据进行连续的风险分析，以了解风险敞口，确定可用的数据保护资源和投资的优先级，并随着大数据的增长制订保护和补救计划。
- 你需要与本地的大数据工具（如 Cloudera Sentry、Amazon Macie 和 Hortonworks Ranger）集成的自动化、集中式大数据隐私工具来简化并促进管理数据访问的过程（如查看、更改和添加访问策略）。
- 你需要大规模、快速、高效的数据保护功能，包括对在生产和数据湖中使用的大数据进行动态屏蔽，对数据湖和数据仓库中静止的大数据进行加密以及对非生产环境中所使用的大数据进行永久性屏蔽。
- 你必须衡量并传达大数据隐私风险指标的状态，这是成功保护敏感信息并支持审计准备工作的关键部分。

# Exercises

**[Ex. 1]** 根据 **Text A** 填空。

1. Big data security is the collective term for _____ and tools used to guard both _____ and _____ from attacks, _____, or other malicious activities that could _____ or negatively affect them.

2. These challenging threats include the theft of information stored online, _____, XSS (Cross Site Scripting) attacks or _____ that could crash a server. Attacks on an organization's big data storage could cause serious financial repercussions such as _____, _____ and _____.

3. Most big data implementations actually distribute _____ across many systems for faster analysis. Distributed processing may mean _____ by any one system, but it means a lot more systems where _____ can crop up.

4. Granular auditing can help determine _____, _____ and _____.

5. The sheer size of a big data installation, terabytes to petabytes large, is _____ for routine security audits. And because most big data platforms are _____, this introduces _____ across multiple nodes and _____.

6. Security tools need to monitor and alert on _____ on the system, database or a web CMS (Content Management System) such as WordPress, and big data security experts must _____ in cleanup and know _____.

7. Centralized key management has been _____ for many years. It is applied in big data environments, especially in those with _____. Best practices include _____, _____, _____, and abstracting key management from key usage.

8. User access control may be _____, but many companies practice minimal control because the management overhead can be _____. Strong user access control requires _____ that automates access based on _____.

9. Physical security systems can deny data center access to _____ or to _____ who _____.

10. Organizations can protect their big data analytics tools_____. One of the most common security tools is _____, a relatively simple tool that can go a long way. Firewalls are effective at _____ that both enters and _____ servers.

**[Ex. 2]** 根据 **Text B** 回答以下问题。

1. What does big data privacy involve?

2. What do you need to do to safeguard big data and ensure it can be used for analytics?

3. What must data owners do in an era of hybrid, multi-cloud computing?

4. What will happen to organizations as they store more types of sensitive data in larger amounts over longer periods of time?

5. What do companies need to respond to these growing demands?

6. What will new big data analytic tools enable organizations to do?

7. What are the four critical data management activities mentioned in the passage?

8. What advantage will companies with a strong and scalable data governance program have when assessing these tasks?

9. What can't big data be? What must it be?

10. What must you do if you want to use that information to develop and implement intelligent big data management policies?

**[Ex. 3]** 词汇英译中

| | |
|---|---|
| 1. big data security | 1. _____ |
| 2. adoption | 2. _____ |
| 3. access control | 3. _____ |
| 4. authenticity | 4. _____ |
| 5. in transit | 5. _____ |
| 6. crash | 6. _____ |
| 7. security service level agreement | 7. _____ |
| 8. damage | 8. _____ |
| 9. sensitive information | 9. _____ |
| 10. delivery | 10. _____ |

**[Ex. 4]** 词汇中译英

| | |
|---|---|
| 1. 人工智能 | 1. _____ |
| 2. 检测，探测 | 2. _____ |
| 3. 社交网络 | 3. _____ |
| 4. 编密码；加密 | 4. _____ |
| 5. 滥用 | 5. _____ |
| 6. 过滤器 | 6. _____ |

7. 攻击，侵袭；损害          7. _____

8. 机密性                    8. _____

9. 入侵                      9. _____

10. 集中式密钥管理           10. _____

[Ex. 5] 短文翻译

## What Is Data Security

### 1. What is data security

Data security is a set of standards and technologies that protect data from intentional or accidental destruction, modification or disclosure. Data security can be applied using a range of techniques, including administrative controls, physical security, logical controls, organizational standards and other safeguarding techniques that limit access to unauthorized or malicious users or processes.

### 2. Why is data security important

All businesses today deal in data to a degree. From the banking giants dealing in massive volumes of personal and financial data to the one-man business storing the contact details of his customers on a mobile phone, data is at play in companies both large and small.

The primary aim of data security is to protect the data that an organization collects, stores, creates, receives or transmits. Compliance is also a major consideration. It doesn't matter which device, technology or process is used to manage, store or collect data, it must be protected. Data breaches can result in litigation cases and huge fines, not to mention damage to an organization's reputation. The importance of shielding data from security threats is more important today than it has ever been.

### 3. Different data security technologies

Data security technology comes in many shapes and forms to protects data from a growing number of threats. Many of these threats are from external sources, but organizations should also focus their efforts on safeguarding their data from the inside. Ways of securing data are as follows.

1) Data encryption: Data encryption applies a code to every individual piece of data and will not grant access to encrypted data without an authorized key being given.

2) Data masking: Masking specific areas of data can protect it from disclosure to external malicious sources, and also internal personnel who could potentially use the data. For example, the first 12 digits of a credit card number may be masked within a database.

3) Data erasure: There are times when data that is no longer active or used needs to be erased from all systems. For example, if a customer has requested for their name to be removed from a mailing list, the details should be deleted permanently.

4) Data resilience: By creating backup copies of data, organizations can recover data should it be erased or corrupted accidentally, or stolen during a data breach.

# Reading Material

## Measures to Protect Big Data Security and Privacy

Big data is nothing new to large organizations, however, it's also becoming popular among smaller and medium sized firms due to cost reduction and provided ease to manage data.

Cloud-based storage has facilitated data mining and collection. However, this big data and cloud storage integration has caused a challenge to privacy and security threats.

The reason for such breaches may be that security applications that are designed to store certain amounts of data cannot deal with the big volumes of data that the aforementioned[1] datasets have. Also, these security technologies are inefficient[2] to manage dynamic data and they can control static data only. Therefore, just a regular security check can not detect security patches[3] for continuous streaming data. For this purpose, you need take some measures while data streaming and big data analysis.

### 1. Protecting transaction logs and data

Data stored in a storage medium[4], such as transaction logs and other sensitive information, may have varying levels, but that's not enough. For instance, the transfer of data between these levels gives the IT manager insight over the data which is being moved. Due to the data continuously increasing, the scalability and availability makes auto-tiering[5] necessary for big data storage management. Yet, new challenges are being posed to big data storage as the auto-tiering method doesn't keep track of[6] data storage location.

### 2. Validation and filtration[7] of end-point[8] inputs

End-point devices are important for maintaining big data. Storing, processing and other necessary tasks are performed with the help of input data, which is provided by end-points. Therefore, an organization should make sure to use authentic and legitimate[9] end-point devices.

### 3. Securing distributed framework calculations and other processes

Computational security and other digital assets[10] in a distributed framework like MapReduce function of Hadoop mostly lack security protections. The two main preventions for it are securing the mappers[11] and protecting the data in the presence of an unauthorized mapper.

---

1  aforementioned [ə,fɔ:'menʃənd] *adj.*上述的；前述的

2  inefficient [,ɪnɪ'fɪʃnt] *adj.*无效率的，无能的

3  patch [pætʃ] *n.*补丁

4  medium ['mi:diəm] *n.*介质

5  auto-tiering ['ɔ:təu taɪərɪŋ] *n.*自动分层

6  keep track of: 跟踪

7  filtration [fɪl'treɪʃn] *n.*过滤；筛选

8  end-point ['endpɔɪnt] *n.*端点，终点

9  legitimate [lɪ'dʒɪtɪmət] *adj.*合法的，合理的

10  digital asset: 数字资产

11  mapper ['mæpə] *n.*映射器

## 4. Securing and protecting data in real time

Due to large amounts of data generation, most organizations are unable to maintain regular checks. However, it is most beneficial to perform security checks and observation in real time or almost in real time.

## 5. Protecting access control method communication and encryption

A secured data storage device is an intelligent step in order to protect the data. Yet, because data storage devices are usually vulnerable[12], it is necessary to encrypt the access control methods as well.

## 6. Determining data provenance

To classify data, it is necessary to be aware of its origin. In order to determine the data origin accurately, authentication, validation and access control could be used.

## 7. Granular auditing

Analyzing different kinds of logs can be advantageous and this information can be helpful in recognizing any kind of cyber attack or malicious activity. Therefore, regular auditing can be beneficial.

## 8. Granular access control

Granular access control of big databases by NoSQL databases or the Hadoop Distributed File System requires a strong authentication process and mandatory[13] access control.

## 9. Protecting databases

Data stores such as NoSQL have many security vulnerabilities, which cause privacy threats. A prominent security flaw[14] is that it is unable to encrypt data during the tagging or logging of data or while distributing it into different groups, when it is streamed or collected.

Organizations must ensure that all big databases are immune to security threats and vulnerabilities. During data collection, all the necessary security protections such as real-time management should be fulfilled. Keeping in mind the huge size of big data, organizations should remember the fact that managing such data could be difficult and requires extraordinary efforts. However, taking all these steps would help maintain consumer privacy.

---

12　vulnerable ['vʌlnərəbl] *adj*.易受攻击的

13　mandatory ['mændətəri] *adj*.强制的；命令的

14　flaw [flɔ:] *n*.错误；缺点

# 附　　录

## 附录 A　大数据专业英语词汇特点及翻译技巧

大数据专业英语的一个显著特点就是新的词汇会不断出现。因此，掌握这些新词汇的构成特点并准确地翻译就十分重要。

**1．大数据专业英语词汇的特点与构造**

大数据专业英语的新词汇中，只有极少数是全新的，绝大多数都是由现有的词汇通过某种方法构造出来的。常见的构造方法如下。

- 新赋意义：给公共英语中的普通词汇在大数据专业语境中赋予新的、有技术特色的意义，使之成为专业术语。如"memory"的常用意思是"记忆"，而在大数据专业英语中给它赋予新的意思"内存"，特指大数据的一种硬件。又如"map"在大数据中的意思是"映射"，library 的意思是"库"，它们都成为技术意味极强的新词。
- 复合构造：由两个或两个以上单词构成一个新词。如 online（在线）、offline（离线）、backlink（反向链接）、database（数据库）、dataset（数据集）及 datastream（数据流）等。
- 派生法构词：在词根前面加前缀或在词根后面加后缀，从而构成一个与原单词意义相近或截然不同的新词，这个方法叫作派生法。例如，可以通过给词根加前缀"un"构造出 unauthorized（*adj*.未经授权的；未经许可的）、undefined（*adj*.未下定义的，不明确的，模糊的）、undiagnosed（*adj*.未诊断的）、unconstructed（*adj*.非结构化的）及 uncouple（*vt*.去耦）等；加"re"可构造出 remodeling（*v*.重构，改造）、reestablish（*vt*.重新建立）、refresh（*v*.刷新，更新）、rekey（*v*.再次输入）、rename（*vt*.更名，改名，重新命名）、restart（*vt.&vi*.重新启动）、restore（*vt*.恢复，修复）及 retransmission（*n*.转播，中继）等；以"er"结尾的词有 filter（*n*.过滤器）、compiler（*n*.编译器）等；以"or"结尾的 calculator（*n*.计算器）、editor（*n*.编辑器）、processor（*n*.处理器）、administrator（管理员）等。

除了以上方法外，还可以用两个或多个单词截开拼接出新词。例如，取出单词 modulate（调制）的前三个字母 mod 与单词 demodulate（解调）的前三个字母 dem，再把两个 d 字母合并，拼接为 modem（调制解调器）。又如 Centrino（迅驰，处理器名）的名字来自 Center（中央）和 Neutrino（微中子）的组合。

- 专业词组：由专业单词组成固定的搭配，表达特定的专业意义。如 variable name（变量名）、virtual machine（虚拟机）、voice recognition（语音识别）、video card（视频卡，显卡）及 open source community（开源社区）、lazy learning algorithm（懒惰学习算法）、deep learning（深度学习）、weak learner（弱分类器）等。

● 大量使用缩略词：大数据专业英语的一个显著特点是用单词首尾字母组成一个新词，在构词法中这叫作首尾字母缩略法。缩略词可以节省大量的篇幅，但也给不熟悉这些缩略词的读者带来了困难。例如，SQL（Structured Query Language，结构化查询语言）、XML（eXtensible Markup Language，可扩展标记语言）、DMP（Data Management Platform，数据管理平台）、DBaaS（DataBase as a Service，数据库即服务）、DDL（Data Definition Language，数据定义语言）、API（Application Programming Interface，应用程序接口）等都属于缩略词。

## 2．大数据专业英语词汇的翻译方法

大数据专业英语词汇汉译的方法多种多样，常用的方法有以下几种。

● 直译（Literal Translation）：所谓直译就是译出原文的字面意义。对有些英语新词语而言，汉语中有与之对应意义的词汇，或者已经约定俗成、可以采用完全直译的方法。例如，把 Microsoft 翻译为"微软"、Oracle 翻译为"甲骨文"。类似的例子有 Natural Language Processing（自然语言处理）、Big Data Firewall（大数据防火墙）、Visualization Platform（可视化平台）、High Level Language（高级语言）、Client Server（客户服务器）、Data Leakage（数据泄露）、Shade（阴影，底纹）等。

在大数据专业英语的汉译方法中，直译法是一种最常用的方法，特别是在翻译复合词时，大都使用直译法。直译法简单易行，便于理解。

● 意译（Liberal Translation）：当一个英语词汇没有完全对应的汉语词汇或者按照英文的字面意思直接翻译不符合汉语的表达方式时，应该使用意译。即用汉语表达出英语词汇的含义，而不是照字面意思"硬译"。例如，mouse 原指"老鼠"，而在大数据学科中却用来指一种指点类输入设备，因此译作"鼠标"。类似的还有 tree（子目录）、laptop（笔记本计算机）、path（路径）、boot（启动）、cache（高速缓冲存储器）、click（点击）、cursor（光标）、debug（调试）、run（运行）及 host（主机）等。

意译时需要翻译者具有良好的大数据专业知识，深刻地理解原文的含义方能准确翻译。

● 音译（Transliteration）：所谓音译，就是按照英语词汇的读音，翻译为相应的汉语词汇。例如，把 Twitter 翻译为"推特"。在最初阶段，这项服务只是用于向好友的手机发送文本信息，是"推送"信息的服务，此处"推特"中的"推"字尤为传神。又如，将 cracker 翻译为"骇客"。因为 cracker 就是从事恶意破解商业软件、恶意入侵他人的网站等事务，让人恐惧。如果翻译为"破解者、攻入者"，就不够形象、生动。

音译大多用于公司名称及产品名称的翻译。例如，Pentium 是 Intel 公司生产的一种大数据微处理芯片，在它刚面世时其运行速度最快，将其翻译为"奔腾"，不仅与原文发音相似，在意义上也体现了该处理器高速快捷的性能，可谓形神俱佳。类似的翻译有 Blog（博客）、Topology（拓扑）、Athlon（速龙，旧译阿斯龙，处理器名）、Core（酷睿，处理器名）、Google（谷歌，公司名）、Cisco（思科，公司名）、Adobe（奥多比，公司名）、Novell（诺威，公司名）等。

● 音意兼译法（Combination of Literal and Liberal Translation）：将音译和意译两种翻译手法结合使用，兼顾语音和词义，翻译时既取其义、亦取其音。例如，把 MicroBlog 翻译为"微博"，这里"微"取"Micro"之义，"博"取"Blog"之音。又如把 Internet

翻译为"因特网",取"Inter"之音、取"net"之义。类似地,还有 Cyberspace(赛博空间)、Ethernet(以太网)及 softcopy(软拷贝)等。

音意兼译法如果运用得当,译文往往新颖,易读易记。

**3. 结束语**

随着信息技术的高速发展,大数据专业英语中的新词汇不断出现,应该依据其构词特点、根据其专业背景知识,运用恰当的翻译方法进行翻译。力求在准确表达其科学性的同时,追求特有的美学特质,并兼顾传播的便捷性。

# 附录 B 单词表

| 单　词 | 音　标 | 意　义 | 单　元 |
| --- | --- | --- | --- |
| abnormality | [ˌæbnɔːˈmæləti] | n.异常;变态;畸形 | 1A |
| absence | [ˈæbsəns] | n.缺乏,缺少 | 2A |
| abstraction | [æbˈstrækʃn] | n.抽象;抽象化 | 4A |
| abuse | [əˈbjuːs] | n.&v.滥用 | 10A |
| accelerate | [əkˈseləreɪt] | v.加快,加速 | 3B |
| access | [ˈækses] | vt.访问,存取(计算机信息) | 2A |
| accessible | [əkˈsesəbl] | adj.可访问的;易接近的 | 3B |
| accommodate | [əˈkɒmədeɪt] | vt.使适应 vi.适应于;作调节 | 9B |
| accomplish | [əˈkʌmplɪʃ] | v.完成,达成 | 7A |
| accumulate | [əˈkjuːmjəleɪt] | v.堆积,积累 | 1A |
| accurate | [ˈækjʊrət] | adj.精确的,准确的 | 1A |
| achieve | [əˈtʃiːv] | vt.取得,获得;实现 | 6A |
| actionable | [ˈækʃənəbl] | adj.可行动的 | 8B |
| adaptation | [ˌædæpˈteɪʃn] | n.改编版;适应 | 8A |
| adhoc | [ˌædˈhɒk] | adj.特别的,特设的;临时的 | 2B |
| adjust | [əˈdʒʌst] | v.调整,调节 | 9B |
| administrative | [ədˈmɪnɪstrətɪv] | adj.管理的,行政的 | 4A |
| administrator | [ədˈmɪnɪstreɪtə] | n.管理者 | 10A |
| adopt | [əˈdɒpt] | v.采用(某方法);采取(某态度) | 5A |
| adoption | [əˈdɒpʃn] | n.采用 | 10A |
| advertisement | [ədˈvɜːtɪsmənt] | n.广告,公告 | 1B |
| afterthought | [ˈɑːftəθɔːt] | n.事后的考虑或想法 | 10B |
| agency | [ˈeɪdʒənsi] | n.服务/代理/经销机构;(政府)专门机构 | 10B |
| aggregate | [ˈæɡrɪɡət] [ˈæɡrɪɡeɪt] | n.合计;聚集体 adj.总计的;聚合的 vt.总计,合计;使聚集,使积聚 | 4B |
| aggregation | [ˌæɡrɪˈɡeɪʃn] | n.聚集;集成;集结 | 5A |

（续）

| 单 词 | 音 标 | 意 义 | 单 元 |
|---|---|---|---|
| agile | ['ædʒaɪl] | adj.敏捷的，灵活的 | 5A |
| alarm | [ə'lɑ:m] | n.警报；闹钟 vt.警告 | 1B |
| algorithm | ['ælgərɪðəm] | n.算法 | 2A |
| alter | ['ɔ:ltə] | vt.改变，更改 | 4A |
| alternative | [ɔ:l'tɜ:nətɪv] | adj.替代的；备选的 n.可供选择的事物 | 3B |
| analogous | [ə'næləgəs] | adj.相似的，可比拟的 | 6A |
| analyst | ['ænəlɪst] | n.分析家，分析师 | 5A |
| analyze | ['ænəlaɪz] | vt.分析；分解 | 1A |
| anomaly | [ə'nɒməli] | n.异常，反常 | 6A |
| anonymize | [ə'nɒnɪmaɪz] | vt.使匿名 | 10B |
| approach | [ə'prəʊtʃ] | n.方法；途径 v.接近，走近，靠近 | 2B |
| approachable | [ə'prəʊtʃəbl] | adj.可亲近的；可接近的 | 7B |
| architecture | ['ɑ:kɪtektʃə] | n.体系结构；（总体、层次）结构 | 7A |
| archival | [ɑ:'kaɪvəl] | adj.档案的 | 3B |
| aspect | ['æspekt] | n.方面；层面；样子；外观 | 6A |
| association | [ə,səʊʃi'eɪʃn] | n.关联；联合，联系 | 6A |
| assumption | [ə'sʌmpʃn] | n.假定，假设 | 6A |
| attack | [ə'tæk] | v.& n.攻击，侵袭；损害 | 10A |
| attain | [ə'teɪn] | v.获得；到达 | 6A |
| attention | [ə'tenʃn] | n.注意，注意力 | 8A |
| attribute | [æ'trɪbju:t] | v.把……归因于 | 2A |
|  | [æ'trɪbju:t] | n.属性；性质；特征 |  |
| audit | ['ɔ:dɪt] | n.&v.审计；检查 | 10A |
| authentication | [ɔ:,θentɪ'keɪʃn] | n.身份验证；认证 | 10A |
| authenticity | [,ɔ:θen'tɪsəti] | n.真实性；可靠性 | 10A |
| authorization | [,ɔ:θəraɪ'zeɪʃn] | n.授权，批准 | 4A |
| automate | ['ɔ:təmeɪt] | v.（使）自动化 | 1A |
| automated | ['ɔ:təmeɪtɪd] | adj.自动化的 | 10B |
| automation | [,ɔ:tə'meɪʃn] | n.自动化（技术），自动操作 | 2A |
| auto-restart | ['ɔ:təʊ,ri:'stɑ:t] | n.自动重启 | 5B |
| available | [ə'veɪləbl] | adj.可获得的；有空的；能找到的 | 1A |
| backlink | ['bæklɪŋk] | n.反向链接 | 6B |
| backup | ['bækʌp] | n.备份 | 3B |
| balance | ['bæləns] | n.平衡 | 3A |
| bandwidth | ['bændwɪdθ] | n.带宽 | 3B |

（续）

| 单　词 | 音　标 | 意　义 | 单元 |
|---|---|---|---|
| batch | [bætʃ] | n.一批 v.分批处理 | 8A |
| behavior | [bɪ'heɪvjə] | n.行为；态度 | 3A |
| benchmark | ['bentʃmɑ:k] | n.基准，参照 | 5B |
| beneficial | [ˌbenɪ'fɪʃl] | adj.有利的，有益的 | 6A |
| bias | ['baɪəs] | n.偏见；偏爱；倾向 | 1A |
| bin | [bɪn] | n.区间；端点，范围 | 8A |
| binary | ['baɪnəri] | adj.二进制的；双重的；二元的 | 2A |
| blackout | ['blækaʊt] | n.停电 | 9B |
| block | [blɒk] | n.块 | 9A |
| blueprint | ['blu:prɪnt] | n.蓝图，设计图；计划大纲 | 2B |
| brand | [brænd] | n.商标，品牌 | 3A |
| breach | [bri:tʃ] | vt.攻破；破坏 n.破坏；破裂 | 1A |
| budget | ['bʌdʒɪt] | n.预算 v.把……编入预算 | 1A |
| built-in | [bɪlt'ɪn] | adj.嵌入的；内置的；固有的 | 5B |
| bundle | ['bʌndl] | n.捆；一批 | 4B |
| calamity | [kə'læməti] | n.灾祸，灾难 | 3B |
| calculation | [ˌkælkjʊ'leɪʃn] | n.计算 | 1B |
| calculus | ['kælkjələs] | n.运算，计算 | 4A |
| camera | ['kæmərə] | n.摄像头；照相机；摄影机 | 1B |
| campaign | [kæm'peɪn] | n.活动；运动 | 3A |
| capacity | [kə'pæsəti] | n.容积；生产量 | 3B |
| capital | ['kæpɪtl] | n.资本；资金 | 3B |
| capture | ['kæptʃə] | vt.&n.捕捉 | 4B |
| cardholder | ['kɑ:dhəʊldə] | n.持有信用卡的人 | 1A |
| career | [kə'rɪə] | n.生涯；经历 adj.就业的 | 8A |
| cashback | ['kæʃbæk] | n.现金返还 | 1B |
| catastrophe | [kə'tæstrəfi] | n.大灾难 | 9B |
| centralized | ['sentrəlaɪzd] | adj.集中的 | 10B |
| chain | [tʃeɪn] | n.链路，链条 | 6A |
| characteristic | [ˌkærəktə'rɪstɪk] | n.特性，特征，特色 adj.特有的；独特的 | 1A |
| chart | [tʃɑ:t] | n.图表 v.记录；绘制……的地图；制订（计划） | 8A |
| class | [klɑ:s] | n.类 | 2A |
| classification | [ˌklæsɪfɪ'keɪʃn] | n.分类，归类 | 2A |
| classifier | ['klæsɪfaɪə] | n.分类器，分类者 | 6B |

（续）

| 单　词 | 音　标 | 意　义 | 单　元 |
|---|---|---|---|
| click | [klɪk] | n.&v.点击 | 3A |
| cluster | ['klʌstə] | n.丛；簇，串；群 vi.丛生；群聚 vt.使聚集 | 6B |
| cluster-based | ['klʌstə beɪst] | adj.基于集群的 | 10A |
| clustering | ['klʌstərɪŋ] | n.聚类 | 6A |
| codify | ['kəʊdɪfaɪ] | vt.编纂，整理；编成法典 | 5A |
| collaboration | [kə,læbə'reɪʃn] | n.合作，协作 | 5B |
| collaborative | [kə'læbərətɪv] | adj.合作的，协作的 | 7B |
| collect | [kə'lekt] | vt.收集 | 3A |
| collectively | [kə'lektɪvli] | adv.全体地，共同地 | 6A |
| column | ['kɒləm] | n.列，纵队 | 2A |
| combine | [kəm'baɪn] | v.使结合；组合 | 5A |
| command | [kə'mɑ:nd] | n.命令 | 4A |
| commoditize | [kə'mɒdɪtaɪz] | v.使商品化 | 9B |
| communicate | [kə'mju:nɪkeɪt] | vt.传达，表达；沟通；传递 | 8A |
| community | [kə'mju:nəti] | n.社团，社区 | 7B |
| comparison | [kəm'pærɪsn] | n.比较，对照 | 5A |
| complementary | [,kɒmplɪ'mentri] | adj.互补的；补充的，补足的 | 7B |
| complex | ['kɒmpleks] | adj.复杂的；复合的 | 5A |
| complexity | [kəm'pleksəti] | n.复杂性 | 2B |
| compliance | [kəm'plaɪəns] | n.服从，听从；承诺 | 3B |
| complicate | ['kɒmplɪkeɪt] | vt.使复杂化；使错综，使混乱 | 2A |
| comply | [kəm'plaɪ] | vi.遵从；依从，顺从 | 3B |
| comprehensive | [,kɒmprɪ'hensɪv] | adj.广泛的；综合的 | 7A |
| comprise | [kəm'praɪz] | vt.包含，包括；由……组成，由……构成 | 9A |
| compromise | ['kɒmprəmaɪz] | v.损害；违背；使陷入危险 n.折中，妥协 | 10A |
| computational | [,kɒmpju'teɪʃənl] | adj.计算的 | 7B |
| concatenation | [kən,kætə'neɪʃn] | n.互相关联的事物 | 5A |
| conceptual | [kən'septʃuəl] | adj.观念的，概念的 | 2B |
| concern | [kən'sɜ:n] | n.令人担心的事；关心（的事）v.影响；涉及；使担忧 | 9A |
| concurrency | [kən'kʌrənsi] | n.并发（性） | 4A |
| concurrency-oriented | [kən'kʌrənsɪ 'ɔ:rɪəntɪd] | adj.面向并发的 | 5B |
| condition | [kən'dɪʃn] | n.状态；环境 vt.制约，限制；使习惯于，使适应 | 6A |
| confidence | ['kɒnfɪdəns] | n.信心，信任 | 6A |
| confident | ['kɒnfɪdənt] | adj.确信的，深信的 | 6A |

（续）

| 单　　词 | 音　　标 | 意　　义 | 单　元 |
|---|---|---|---|
| confidentiality | [ˌkɒnfɪˌdenʃiˈæləti] | n.机密性 | 10A |
| confuse | [kənˈfjuːz] | vt.使混乱；使困惑；使难理解 | 8A |
| congestion | [kənˈdʒestʃən] | n.拥挤，阻塞 | 3B |
| connection | [kəˈnekʃn] | n.连接；联系 | 3B |
| connector | [kəˈnektə] | n.连接器 | 8B |
| consequence | [ˈkɒnsɪkwəns] | n.结果；重要性 | 10A |
| consistency | [kənˈsɪstənsi] | n.一致性 | 2B |
| consistent | [kənˈsɪstənt] | adj.一致的；连续的 | 2A |
| consolidate | [kənˈsɒlɪdeɪt] | v.统一；合并；联合 | 8B |
| constraint | [kənˈstreɪnt] | n.限制；约束 | 3B |
| construct | [kənˈstrʌkt] | vt.构建，构造 | 6A |
| construction | [kənˈstrʌkʃn] | n.建造；建造物 | 2B |
| consumption | [kənˈsʌmpʃn] | n.消费 | 1A |
| contender | [kənˈtendə] | n.（冠军）争夺者，竞争者 | 7A |
| continuity | [ˌkɒntɪˈnjuːəti] | n.连续性，连接 | 3B |
| continuous | [kənˈtɪnjuəs] | adj.连续的，不断的 | 1A |
| continuously | [kənˈtɪnjuəsli] | adv.连续不断地 | 3A |
| contribution | [ˌkɒntrɪˈbjuːʃn] | n.贡献 | 1B |
| control | [kənˈtrəʊl] | vt.控制；管理 | 4A |
| conveniently | [kənˈviːniəntli] | adv.方便地，便利地 | 8B |
| conventional | [kənˈvenʃənl] | adj.传统的；依照惯例的，平常的 | 1A |
| convergence | [kənˈvɜːdʒəns] | n.集合；会聚 | 9A |
| conversation | [ˌkɒnvəˈseɪʃn] | n.交谈，会话；人机对话 | 8B |
| conversion | [kənˈvɜːʃn] | n.变换，转变 | 5A |
| copy | [ˈkɒpi] | v.复制 | 5A |
| correlation | [ˌkɒrəˈleɪʃn] | n.相互关系；相关性 | 6B |
| corresponding | [ˌkɒrəˈspɒndɪŋ] | adj.相当的，对应的；符合的 | 6A |
| corrupt | [kəˈrʌpt] | adj.损坏的，（文献等）错误百出的 | 5A |
| cost-effective | [kɒst ɪˈfektɪv] | adj.有成本效益的，划算的 | 3B |
| counter | [ˈkaʊntə] | n.计数器 | 3A |
| counterpart | [ˈkaʊntəpɑːt] | n.副本；相对物；极相似的人或物 | 9B |
| crash | [kræʃ] | n.崩溃；碰撞 v.（使）崩溃，（使）瘫痪 | 10A |
| create | [kriˈeɪt] | vt.建立 | 4A |
| critical | [ˈkrɪtɪkl] | adj.关键的；严重的；极重要的 | 5A |
| critically | [ˈkrɪtɪkli] | adv.危急地；严重地 | 10A |

（续）

| 单 词 | 音 标 | 意 义 | 单 元 |
|---|---|---|---|
| crossover | ['krɒsəʊvə] | n.杂交，交换 | 7A |
| cross-platform | [krɒs 'plætfɔ:m] | adj.跨平台的 | 5B |
| curiosity | [ˌkjʊəri'ɒsəti] | n.好奇，求知欲 | 10A |
| cutting-edge | [ˌkʌtɪŋ'edʒ] | adj.前沿的 | 4B |
| cyber | ['saɪbə] | adj.计算机（网络）的 | 10A |
| damage | ['dæmɪdʒ] | v.&n.损坏，伤害 | 10A |
| dashboard | ['dæʃbɔ:d] | n.仪表板；仪表盘 | 5B |
| database | ['deɪtəbeɪs] | n.数据库 | 4A |
| data-gathering | ['deɪtə 'gæðərɪŋ] | n.数据收集 | 3A |
| data-intensive | ['deɪtə ɪn'tensɪv] | adj.数据密集的 | 9A |
| decision | [dɪ'sɪʒn] | n.决定 | 4B |
| declarative | [dɪ'klærətɪv] | adj.声明的，陈述的 | 4A |
| define | [dɪ'faɪn] | vi.下定义，构成释义 n.定义 | 2A |
| definitely | ['defɪnətli] | adv.确定地；明显地；明确地 | 7A |
| degrade | [dɪ'greɪd] | vt.降低，降级 | 5A |
| delete | [dɪ'li:t] | v.删除 | 4A |
| deliver | [dɪ'lɪvə] | vt.交付；发表 | 3B |
| delivery | [dɪ'lɪvəri] | n.交付 | 10A |
| demand | [dɪ'mɑ:nd] | v.&n.需求，需要 | 3B |
| demographic | [ˌdemə'græfɪk] | adj.人口统计学的 | 6A |
| demonstrate | ['demənstreɪt] | vt.证明，证实；显示，展示 | 3A |
| deny | [dɪ'naɪ] | v.否认；拒绝 | 10A |
| depict | [dɪ'pɪkt] | vt.描绘，描述 | 2B |
| deploy | [dɪ'plɔɪ] | v.部署 | 9A |
| deployable | [dɪ'plɔɪeɪbl] | adj.可部署的 | 7A |
| deployment | [dɪ'plɔɪmənt] | n.部署；调度 | 6A |
| derive | [dɪ'raɪv] | v.得到，导出，源于 | 6A |
| describe | [dɪ'skraɪb] | vt.叙述；描绘 | 2A |
| destroy | [dɪ'strɔɪ] | vt.破坏，摧毁，使失败 | 9B |
| detection | [dɪ'tekʃn] | n.检测，探测 | 10A |
| deterrent | [dɪ'terənt] | n.制止物；威慑物 | 9B |
| developer | [dɪ'veləpə] | n.开发者 | 4A |
| device | [dɪ'vaɪs] | n.装置，设备 | 2A |
| diagram | ['daɪəgræm] | n.图表；图解；示意图 | 2B |
| dialect | ['daɪəlekt] | n.方言，专业用语 | 8B |

（续）

| 单　词 | 音　标 | 意　义 | 单　元 |
|---|---|---|---|
| dimensional | [dɪ'menʃənəl] | adj.维度的，维数的 | 2B |
| directory | [də'rektəri] | n.（计算机文件或程序的）目录 | 9A |
| disaster | [dɪ'zɑ:stə] | n.灾难 | 3B |
| discern | [dɪ'sɜ:n] | v.辨别，分清 | 8A |
| discomfort | [dɪs'kʌmfət] | n.不舒适，不舒服；不安 | 3B |
| discover | [dɪ'skʌvə] | vt.发现；获得知识 | 6A |
| disease | [dɪ'zi:z] | n.疾病；弊端 | 1B |
| disparate | ['dɪspərət] | adj.完全不同的，根本不同的 | 4B |
| distinct | [dɪ'stɪŋkt] | adj.明显的，清楚的 | 9A |
| distribute | [dɪ'strɪbju:t] | vt.分布，分配；散发，分发 | 4A |
| document-oriented | ['dɒkjʊmənt 'ɔ:rɪəntɪd] | adj.面向文档的 | 5B |
| dominant | ['dɒmɪnənt] | adj.占优势的；统治的，支配的 | 4A |
| dramatic | [drə'mætɪk] | adj.戏剧性的；引人注目的；突然的 | 4A |
| dramatically | [drə'mætɪk(ə)li] | adv.戏剧性地，引人注目地；显著地 | 7B |
| duplicate | ['dju:plɪkeɪt] | v.复制 adj.复制的；副本的 n.复制品；副本 | 5A |
| duplication | [ˌdju:plɪ'keɪʃn] | n.复制；重复；成倍 | 9A |
| e-commerce | [i:'kɒmɜ:s] | n.电子商务 | 4A |
| ecosystem | ['i:kəʊsɪstəm] | n.生态系统 | 7A |
| effortlessly | ['efətləsli] | adv.不作努力地，不费力地 | 7A |
| elastically | [ɪ'læstɪkli] | adv.有弹性地，伸缩自如地 | 9A |
| elasticity | [ˌi:læ'stɪsəti] | n.弹性；灵活性；伸缩性 | 3B |
| electronically | [ɪˌlek'trɒnɪkli] | adv.电子地 | 4A |
| element | ['elɪmənt] | n.元素；要素；原理 | 2A |
| eliminate | [ɪ'lɪmɪneɪt] | vt.排除，消除 | 4B |
| email | ['i:meɪl] | n.电子邮件 vt.给……发电子邮件 | 1A |
| emphasize | ['emfəsaɪz] | v.（使）突出/明显；强调；重视 | 8A |
| empower | [ɪm'paʊə] | vt.授权，准许；使能够 | 5B |
| empty | ['empti] | adj.空的 | 5A |
| encompass | [ɪn'kʌmpəs] | vt.包含或包括某事物；完成 | 4B |
| encounter | [ɪn'kaʊntə] | vt.不期而遇；遭遇 | 8A |
| encourage | [ɪn'kʌrɪdʒ] | v.鼓励；促进 | 10B |
| encryption | [ɪn'krɪpʃn] | n.编密码；加密 | 10A |
| enforcement | [ɪn'fɔ:smənt] | n.强制，实施，执行 | 4A |
| engine | ['endʒɪn] | n.引擎，发动机 | 4A |
| enhance | [ɪn'hɑ:ns] | vt.提高，增加；加强 | 1A |

| 单　　词 | 音　　标 | 意　　义 | 单　元 |
|---|---|---|---|
| enhancement | [ɪn'hɑ:nsmənt] | n.增强；提高；改善 | 7B |
| enormous | [ɪ'nɔ:məs] | adj.巨大的，庞大的 | 1A |
| ensemble | [ɒn'sɒmbl] | n.集成；全体 | 6B |
| ensure | [ɪn'ʃʊə] | vt.确保 | 2A |
| entity | ['entəti] | n.实体 | 2A |
| environmental | [ɪn,vaɪrən'mentl] | adj.自然环境的 | 1B |
| equation | [ɪ'kweɪʒn] | n.方程式；等式 | 6B |
| erase | [ɪ'reɪz] | vt.擦掉；抹去；清除 | 5A |
| espouse | [ɪ'spauz] | vt.支持；拥护；赞助 | 9A |
| essential | [ɪ'senʃl] | adj.基本的；必要的；本质的 | 3A |
| establish | [ɪ'stæblɪʃ] | vt.建立，创建，确立 | 3A |
| estimate | ['estɪmət]<br>['estɪmeɪt] | n.估价，估算<br>v.估价，估算 | 6B |
| evaluate | [ɪ'væljueɪt] | v.估计 | 8B |
| evidence | ['evɪdəns] | n.证据；迹象 vt.显示；表明；证实 | 4B |
| exaggerate | [ɪg'zædʒəreɪt] | v.夸张；夸大 | 6B |
| excellent | ['eksələnt] | adj.卓越的；杰出的；优秀的 | 3A |
| exceptional | [ɪk'sepʃənl] | adj.杰出的，非凡的 | 10A |
| exchange | [ɪks'tʃeɪndʒ] | n.&v.交换，互换 | 2A |
| executable | [ɪg'zekjətəbl] | adj.可执行的；实行的 | 2A |
| execution | [,eksɪ'kju:ʃn] | n.实行，履行，执行 | 6A |
| expert | ['ekspɜ:t] | n.专家，能手；权威；行家 adj.专业的 | 1A |
| expertise | [,ekspɜ:'ti:z] | n.专门知识或技能；专家的意见 | 1A |
| exploratory | [ɪk'splɒrətri] | adj.探索的 | 6A |
| explore | [ɪk'splɔ:] | vt.探索，探究 | 1A |
| exponential | [,ekspə'nenʃl] | adj.指数的，幂数的 | 6A |
| exponentially | [,ekspə'nenʃəli] | adv.以指数方式 | 1A |
| export | [ɪk'spɔ:t] | v.（计算机）导出 | 8B |
| exposure | [ɪk'spəʊʒə] | n.暴露 | 10A |
| extra | ['ekstrə] | adj.额外的，补充的，附加的 | 5A |
| eye-catching | ['aɪ ,kætʃɪŋ] | adj.引人注目的；显著的 | 5B |
| facilitate | [fə'sɪlɪteɪt] | vt.促进，助长；使容易 | 4A |
| factor | ['fæktə] | n.因素 | 6A |
| failure | ['feɪljə] | n.失败 | 5A |
| fault-tolerant | ['fɔ:lt 'tɒlərənt] | adj.容错的 | 5B |

（续）

| 单 词 | 音 标 | 意 义 | 单 元 |
|---|---|---|---|
| favourite | ['feɪvərɪt] | adj.特别受喜爱的 n.特别喜爱的人（或物） | 1B |
| field | [fi:ld] | n.字段，域 | 2A |
| file | [faɪl] | n.文件 | 2A |
| filename | ['faɪlneɪm] | n.文件名 | 9A |
| filter | ['fɪltə] | n.过滤器 | 10A |
| filtering | ['fɪltərɪŋ] | v.过滤 | 5A |
| fine | [faɪn] | n.罚款 v.对……处以罚金 | 10A |
| fine-grained | ['faɪn'greɪnd] | adj.精确的 | 5B |
| finite | ['faɪnaɪt] | adj.有限的 | 3B |
| firewall | ['faɪəwɔ:l] | n.防火墙 | 9B |
| flash | [flæʃ] | n.闪存 | 10A |
| flaw | [flɔ:] | n.瑕疵，缺点 | 7A |
| flexible | ['fleksəbl] | adj.灵活的 | 2A |
| follower | ['fɒləʊə] | n.关注者，追随者，拥护者 | 3A |
| format | ['fɔ:mæt] | n.格式 vt.使格式化 | 2A |
| formula | ['fɔ:mjələ] | n.公式，准则 | 8A |
| formulate | ['fɔ:mjuleɪt] | vt.构想出，规划；用公式表示 | 6A |
| forum | ['fɔ:rəm] | n.论坛 | 8B |
| foundation | [faʊn'deɪʃn] | n.基础 | 2A |
| framework | ['freɪmwɜ:k] | n.架构，框架；（体系的）结构 | 1A |
| framing | ['freɪmɪŋ] | n.构架；框架 | 7A |
| fraud | [frɔ:d] | n.欺诈；骗子 | 1A |
| fraudulent | ['frɔ:djələnt] | adj.欺骗的，不诚实的 | 1A |
| frequency | ['fri:kwənsi] | n.频繁性，频率 | 8A |
| frustrated | [frʌ'streɪtɪd] | adj.挫败的，失意的，泄气的 | 7A |
| fuel | ['fju:əl] | n.燃料 | 1B |
| functionality | [ˌfʌŋkʃə'næləti] | n.功能性 | 4B |
| fusion | ['fju:ʒn] | n.融合 | 5B |
| gather | ['gæðə] | vt.收集；采集 | 4B |
| generalizable | ['dʒenərəlaɪzəbl] | adj.可泛化的，可推广的 | 6A |
| generalization | [ˌdʒenrəlaɪ'zeɪʃn] | n.泛化，概括 | 6A |
| generalize | ['dʒenrəlaɪz] | v.泛化；推广 | 6A |
| generate | ['dʒenəreɪt] | vt.形成，造成；引起 | 6B |
| genre | ['ʒɒnrə] | n.类型，种类 | 6A |
| geographical | [ˌdʒi:ə'græfɪkl] | adj.地理的 | 10A |

| 单　词 | 音　标 | 意　义 | 单　元 |
|---|---|---|---|
| geospatial | [ˌdʒiːəʊ'speɪʃəl] | adj.地理空间的 | 5B |
| governance | ['gʌvənəns] | n.治理，管理 | 10B |
| granularity | [ˌgrænjə'lærəti] | n.间隔尺寸，粒度 | 5A |
| graph | [grɑːf] | n.图表，曲线图 | 2A |
| groundbreaking | ['graʊndbreɪkɪŋ] | adj.开创性的，突破性的 | 4A |
| guarantee | [ˌgærən'tiː] | v.担保；确保 n.保证；保修单 | 8B |
| guard | [gɑːd] | v.保护；守卫 | 10A |
| habit | ['hæbɪt] | n.习惯，习性 | 1B |
| hacker | ['hækə] | n.黑客 | 4A |
| handle | ['hændl] | vi.处理；操作，操控 n.把手；方法；提手 | 1A |
| handy | ['hændi] | adj.方便的；手边的；便于使用的 | 6A |
| hardcore | ['hɑːdkɔː] | n.核心部分 | 7A |
| harm | [hɑːm] | v.&n.伤害，危害 | 10A |
| heterogeneity | [ˌhetərədʒə'niːəti] | n.异质性，不均匀性 | 2A |
| hierarchical | [ˌhaɪə'rɑːkɪkl] | adj.分层的 | 2A |
| high-performance | [haɪ pə'fɔːməns] | adj.高性能的 | 9A |
| histogram | ['hɪstəgræm] | n.柱状图；直方图 | 8A |
| host | [həʊst] | n.主机 v.托管 | 3B |
| housekeeper | ['haʊskiːpə] | n.管家；主妇 | 1B |
| hybrid | ['haɪbrɪd] | n.混合物 | 10B |
| hyperplane | ['haɪpəpleɪn] | n.超平面 | 6B |
| identifiable | [aɪˌdentɪ'faɪəbl] | adj.可识别的，可辨认的 | 2A |
| identity | [aɪ'dentəti] | n.身份；个性 | 10B |
| ignore | [ɪg'nɔː] | v.忽略，不理 | 10A |
| illustrate | ['ɪləstreɪt] | vt.给……加插图；（用示例、图画等）说明 | 8A |
| implement | ['ɪmplɪment] | vt.实施，执行；实现 n.工具，器械；手段 | 2B |
| implementation | [ˌɪmplɪmen'teɪʃn] | n.实施 | 6B |
| implicit | [ɪm'plɪsɪt] | adj.无疑问的，绝对的；成为一部分的 | 2A |
| imply | [ɪm'plaɪ] | v.暗示，意味，隐含 | 6A |
| improvement | [ɪm'pruːvmənt] | n.改进之处，改善 | 1A |
| inaccuracy | [ɪn'ækjərəsi] | n.不准确，误差 | 10B |
| inception | [ɪn'sepʃn] | n.开始，开端，初期 | 2B |
| independence | [ˌɪndɪ'pendəns] | n.独立，自主 | 4A |
| index | ['ɪndeks] | n.索引；指数 vt.给……编索引；把……编入索引 | 2A |
| indispensable | [ˌɪndɪ'spensəbl] | adj.不可缺少的，绝对必要的 | 7B |

（续）

| 单　词 | 音　标 | 意　义 | 单　元 |
|---|---|---|---|
| inevitable | [ɪnˈevɪtəbl] | adj.不可避免的，必然发生的 | 3B |
| inference | [ˈɪnfərəns] | n.推理，推断，推论 | 10B |
| influencer | [ˈɪnfluənsə] | n.意见领袖；有影响力的人 | 8A |
| ingest | [ɪnˈdʒest] | vt.获取（某事物）；吸收 | 5B |
| ingress | [ˈɪngres] | n.进入，进入权 | 10A |
| inherent | [ɪnˈhɪərənt] | adj.固有的，内在的 | 9A |
| inherit | [ɪnˈherɪt] | v.继承 | 2B |
| initiative | [ɪˈnɪʃətɪv] | n.主动性；主动权；倡议 | 6A |
| innovative | [ˈɪnəveɪtɪv] | adj.革新的，创新的 | 1A |
| insert | [ɪnˈsɜːt] | vt.插入；嵌入 | 4A |
| insertion | [ɪnˈsɜːʃn] | n.插入 | 5B |
| insight | [ˈɪnsaɪt] | n.见解；洞察力；领悟 | 8B |
| insightful | [ˈɪnsaɪtfʊl] | adj.富有洞察力的，有深刻见解的 | 5A |
| instance | [ˈɪnstəns] | n.实例 v.举……为例 | 2B |
| instant | [ˈɪnstənt] | n.瞬间，顷刻 adj.立即的 | 4A |
| instruction | [ɪnˈstrʌkʃn] | n.指令 | 4A |
| insurance | [ɪnˈʃʊərəns] | n.预防措施；保险，保险业；保险费 | 3B |
| integration | [ˌɪntɪˈgreɪʃn] | n.整合；一体化 | 1A |
| integrator | [ˈɪntɪgreɪtə] | n.综合者；积分器（仪、机、元件、电路、装置） | 9A |
| interaction | [ˌɪntərˈækʃn] | n.互相影响；互动 | 3A |
| interactive | [ˌɪntərˈæktɪv] | adj.交互式的；互动的 | 7A |
| interconnectivity | [ɪntəkənekˈtəvɪti] | n.互联性；互联互通；连通性 | 7B |
| interface | [ˈɪntəfeɪs] | n.界面；接口 | 4A |
| intermediate | [ˌɪntəˈmiːdiət] | adj.中间的，中级的 n.中间物 | 5A |
| internal | [ɪnˈtɜːnl] | adj.内部的 | 3B |
| internet | [ˈɪntənet] | n.互联网 | 3B |
| interoperability | [ˌɪntərˌɒpərəˈbɪləti] | n.互用性，协同工作的能力 | 7A |
| interpret | [ɪnˈtɜːprət] | v.解释 | 6B |
| interpretability | [ɪnˈtɜːprɪtəbɪlɪti] | n.可解释性；解释能力；可解读性 | 5A |
| interpretation | [ɪnˌtɜːprɪˈteɪʃn] | n.解释，说明 | 1B |
| interruption | [ˌɪntəˈrʌpʃn] | n.中断，打断 | 3B |
| interval | [ˈɪntəvl] | n.（时间上的）间隔，区间 | 8A |
| introduce | [ˌɪntrəˈdjuːs] | vt.引进；提出；介绍 | 10A |
| intrusion | [ɪnˈtruːʒn] | n.入侵 | 10A |

（续）

| 单　词 | 音　标 | 意　义 | 单　元 |
|---|---|---|---|
| inventive | [ɪn'ventɪv] | adj.发明的；有创造力的 | 4A |
| investment | [ɪn'vestmənt] | n.投资；（时间/精力的）投入 | 10B |
| irregular | [ɪ'regjələ] | adj.不规则的；无规律的；不合规范的 | 2A |
| irrespective | [ɪrɪ'spektɪv] | adj.不考虑的，不顾的；无关的 | 5A |
| iteration | [ˌɪtə'reɪʃn] | n.迭代；循环 | 6B |
| latency | ['leɪtənsi] | n.延迟；潜伏 | 3B |
| layout | ['leɪaut] | n.布局，安排，设计；层 | 5B |
| leakage | ['li:kɪdʒ] | n.漏出；泄露 | 6A |
| legacy | ['legəsi] | n.遗产 | 2B |
| leverage | ['li:vərɪdʒ] | v.利用；发挥杠杆作用 n.杠杆作用；优势，力量 | 9A |
| library | ['laɪbrəri] | n.库 | 7A |
| license | ['laɪsns] | n.执照，许可证；特许 vt.发许可证 | 4B |
| lifecycle | ['laɪf,saɪkl] | n.生命周期 | 2B |
| likelihood | ['laɪklɪhʊd] | n.可能，可能性；[数]似然，似真 | 6A |
| link | [lɪŋk] | v.连接 n.联系，关系 | 3A |
| load | [ləʊd] | v.加载；装载 n.负荷；装载；工作量 | 4B |
| log | [lɒg] | n.记录；日志 | 2A |
| lookup | ['lʊkʌp] | v.查找；查表 | 5A |
| mainframe | ['meɪnfreɪm] | n.主机；大型机 | 4B |
| mainstream | ['meɪnstri:m] | n.主流 adj.主流的 | 2A |
| maintenance | ['meɪntənəns] | n.维护；维修 | 2B |
| malicious | [mə'lɪʃəs] | adj.恶意的；蓄意的；预谋的 | 10A |
| malware | ['mælweə] | n.恶意软件，流氓软件 | 10A |
| mandate | ['mændeɪt] | n.授权；委任 | 10B |
| manipulation | [mə,nɪpjʊ'leɪʃn] | n.操作；控制 | 4A |
| manual | ['mænjuəl] | adj.手动的，手工的 n.使用手册，说明书 | 8A |
| map | [mæp] | v.映射 | 5A |
| massive | ['mæsɪv] | adj.大量的；大规模的 | 1A |
| mathematical | [ˌmæθə'mætɪkl] | adj.数学的 | 6A |
| matrix | ['meɪtrɪks] | n.矩阵；模型 | 8A |
| mature | [mə'tʃʊə] | adj.成熟的；到期的 | 10A |
| measure | ['meʒə] | v.衡量；测量；估量 n.尺度；度量单位 | 8B |
| mention | ['menʃn] | vt.提到，说起 | 3A |
| merge | [mɜ:dʒ] | v.（使）混合；相融；融入 | 5A |
| mesh | [meʃ] | v.（使）吻合；相配，匹配；紧密配合 | 8B |

（续）

| 单 词 | 音 标 | 意 义 | 单 元 |
|---|---|---|---|
| metadata | ['metədeɪtə] | n.元数据 | 2A |
| metrics | ['metrɪks] | n.指标；度量 | 7B |
| middleware | ['mɪdlweə] | n.中间设备，中间件 | 9A |
| migrate | [maɪ'greɪt] | vi.迁移，移往；移动 | 9B |
| migration | [maɪ'greɪʃn] | n.迁移 | 9A |
| minimize | ['mɪnɪmaɪz] | vt.把……减至最低数量（程度） | 10B |
| modify | ['mɒdɪfaɪ] | v.修改，改变 | 4A |
| moisture | ['mɔɪstʃə] | n.水分；湿气 | 1B |
| monitoring | ['mɒnɪtərɪŋ] | n.监视；控制；监测 | 3A |
| motivation | [,məʊtɪ'veɪʃn] | n.动力；积极性 | 10A |
| multi-cloud | ['mʌltɪ klaʊd] | n.多云 | 10B |
| multi-dimension | ['mʌltɪ daɪ'menʃn] | n.多维 | 9A |
| multi-faceted | ['mʌltɪ 'fæsitid] | adj.多方面的；多元化的 | 10A |
| multimedia | [,mʌltɪ'miːdɪə] | adj.多媒体的 | 5B |
| multimodel | ['mʌltɪ'mɒdl] | n.多模式 | 4A |
| multiple | ['mʌltɪpl] | adj.多重的；多个的；多功能的 | 10A |
| native | ['neɪtɪv] | adj.本地的 | 10B |
| negatively | ['negətɪvli] | adv.负面地，否定地，消极地 | 9B |
| neighborhood | ['neɪbəhʊd] | n.地区；某地区的人 | 6A |
| network | ['netwɜːk] | n.（计算机）网络 | 2A |
| neuron | ['njʊərɒn] | n.神经元；神经细胞 | 6A |
| node | [nəʊd] | n.节点 | 2B |
| noise | [nɔɪz] | n.噪声 | 1A |
| normalization | [,nɔːməlaɪ'zeɪʃn] | n.归一化 | 6A |
| normalize | ['nɔːməlaɪz] | vt.使规范化；使正常化；使标准化 | 4B |
| notification | [,nəʊtɪfɪ'keɪʃn] | n.通知；布告；公布 | 5A |
| nuance | ['njuː ɑːns] | n.细微差别 | 7B |
| numerous | ['njuːmərəs] | adj.很多的，许多的 | 9A |
| object | ['ɒbdʒɪkt] | n.物体；目标；对象 | 2A |
| object-based | ['ɒbdʒɪkt beɪst] | adj.基于对象的 | 9A |
| object-oriented | ['ɒbdʒɪkt 'ɔːrɪəntɪd] | adj.面向对象的 | 4A |
| obligation | [,ɒblɪ'geɪʃn] | n.义务，责任；债务 | 9B |
| obstacle | ['ɒbstəkl] | n.障碍（物） | 1B |
| obtain | [əb'teɪn] | v.得到；存在 | 8B |
| obvious | ['ɒbvɪəs] | adj.明显的；当然的 | 10A |

大数据专业英语教程

（续）

| 单　词 | 音　标 | 意　义 | 单　元 |
|---|---|---|---|
| occasional | [ə'keɪʒənl] | adj.偶尔的，临时的 | 3B |
| off-site | ['ɔːf saɪt] | adj.非现场的；站外的；异地的 | 3B |
| on-demand | [ɒn dɪ'mɑːnd] | adj.按需的 | 9A |
| online | [,ɒn'laɪn] | adj.在线的；联网的；联机的 | 3A |
| on-premise | [ɒn'premɪs] | adj.本地的；预置的 | 10A |
| operation | [,ɒpə'reɪʃn] | n.操作；运算 | 2A |
| opinion | [ə'pɪnjən] | n.意见，主张 | 8B |
| opportunity | [,ɒpə'tjuːnəti] | n.机会；时机 | 8A |
| optimization | [,ɒptɪmaɪ'zeɪʃn] | n.最佳化，最优化 | 7B |
| optimize | ['ɒptɪmaɪz] | vt.使最优化 | 4A |
| option | ['ɒpʃn] | n.选择；选择权 | 9B |
| organizational | [,ɔːgənaɪ'zeɪʃənl] | adj.组织的 | 2A |
| outlier | ['aʊtlaɪə] | n.离群值；异常值 | 6A |
| overcome | [,əʊvə'kʌm] | v.克服；战胜 | 10A |
| overflow | [,əʊvə'fləʊ] | v.溢出 | 9B |
| overhead | ['əʊvəhed] | n.经常性支出，运营费用 | 10A |
| overload | [,əʊvə'ləʊd] | vt.使负担太重；使超载；超过负荷 | 1A |
| packet | ['pækɪt] | n.信息包 vt.包装，打包 | 2A |
| painstaking | ['peɪnzteɪkɪŋ] | adj.艰苦的，辛苦的 n.辛苦，勤勉 | 2B |
| parallel | ['pærəlel] | adj.并行的；平行的 | 5B |
| parallelism | ['pærəlelɪzəm] | n.平行；对应；类似 | 7B |
| parameter | [pə'ræmɪtə] | n.参数 | 1B |
| partial | ['pɑːʃl] | adj.部分的 | 2A |
| participant | [pɑː'tɪsɪpənt] | n.参加者，参与者 adj.参与的 | 3A |
| particular | [pə'tɪkjələ] | adj.特别的；详细的；独有的 n.细节；详情 | 1B |
| path | [pɑːθ] | n.路径，路线 | 2B |
| pattern | ['pætn] | n.模式 | 4B |
| percentage | [pə'sentɪdʒ] | n.百分比 | 8A |
| period | ['pɪəriəd] | n.一段时间；阶段；时期 | 10B |
| permission | [pə'mɪʃn] | n.准许；许可证 | 10A |
| persistence | [pə'sɪstəns] | n.持久性，持久化 | 7B |
| persistent | [pə'sɪstənt] | adj.持续的；坚持不懈的；持久的 | 10B |
| petabyte | ['petəbaɪt] | n.拍字节，缩写为 PB，1PB=1024TB=$2^{50}$B | 10A |
| phase | [feɪz] | n.阶段 | 2B |
| pick | [pɪk] | v.挑选 | 9B |

（续）

| 单 词 | 音 标 | 意 义 | 单 元 |
|---|---|---|---|
| policy-based | ['pɒləsɪ beɪst] | adj.基于策略的 | 10A |
| pool | [pu:l] | n.池 | 9B |
| population | [,pɒpjʊ'leɪʃn] | n.人口 | 6A |
| pop-up | ['pɒp ʌp] | adj.弹出的 | 3A |
| portability | [,pɔ:tə'bɪləti] | n.可移植性，可迁移性 | 9A |
| portable | ['pɔ:təbl] | adj.手提的；轻便的 | 2A |
| possibility | [,pɒsə'bɪləti] | n.可能性 | 1A |
| potential | [pə'tenʃl] | adj.潜在的 n.潜力；可能性 | 9B |
| potentially | [pə'tenʃəli] | adv.潜在地，可能地 | 1A |
| pre-built | ['pri:bɪlt] | adj.预建的 | 8B |
| predefine | ['pri:dɪ'faɪn] | vt.预定义；预先确定 | 4B |
| pre-defined | [pri dɪ'faɪnd] | adj.预先定义的 | 2A |
| prediction | [prɪ'dɪkʃn] | n.预测，预报；预言 | 6A |
| preference | ['prefrəns] | n.偏爱；优先权 | 10B |
| preparation | [,prepə'reɪʃn] | n.准备，预备 | 6A |
| prepare | [prɪ'peə] | vt.准备；预备 | 4B |
| presentation | [,prezn'teɪʃn] | n.幻灯片 | 2A |
| pressure | ['preʃə] | n.压力 | 10B |
| prevailing | [prɪ'veɪlɪŋ] | adj.占优势的；主要的；普遍的 | 5A |
| prevalent | ['prevələnt] | adj.流行的；普遍的 | 2B |
| preventative | [prɪ'ventətɪv] | adj.预防性的 | 4A |
| prevention | [prɪ'venʃn] | n.预防；阻止 | 10A |
| prioritize | [praɪ'ɒrətaɪz] | vt.按重要性排列，划分优先顺序；优先处理 | 10B |
| priority | [praɪ'ɒrəti] | n.优先，优先权；重点 | 9B |
| privacy | ['prɪvəsi] | n.隐私，秘密 | 10A |
| privilege | ['prɪvəlɪdʒ] | n.特权；优惠 | 10A |
| privileged | ['prɪvəlɪdʒd] | adj.享有特权的；特许的，专用的 | 10A |
| probability | [,prɒbə'bɪləti] | n.可能性；概率 | 6A |
| procedural | [prə'si:dʒərəl] | adj.程序的，过程的 | 4A |
| process | ['prəʊses] | vt.加工；处理 n.步骤，程序；过程 | 1A |
| procurement | [prə'kjʊəmənt] | n.采购；获得，取得 | 3B |
| production | [prə'dʌkʃn] | n.生产，制作；产品；产量 | 1A |
| proficiency | [prə'fɪʃnsi] | n.熟练，精通，娴熟 | 1A |
| proficient | [prə'fɪʃnt] | adj.精通的，熟练的 | 10A |
| profit | ['prɒfɪt] | n.利润；红利 vi.有益；获利 | 6A |

（续）

| 单　词 | 音　标 | 意　义 | 单　元 |
|---|---|---|---|
| program | ['prəugræm] | n.程序；计划，安排 v.给……编写程序；为……制订计划 | 1A |
| programmability | [,prəugræmə'bılıti] | n.可编程性 | 7B |
| programmatically | [,prəugrə'mætıkli] | adv.编程地 | 5B |
| programmer | ['prəugræmə] | n.程序员，程序设计者 | 2B |
| project | ['prɒdʒekt] | n.项目；方案；工程，计划 | 5A |
| proliferation | [prə,lıfə'reıʃn] | n.增长 | 10B |
| promote | [prə'məut] | vt.促进，推进 | 1A |
| promotion | [prə'məuʃn] | n.促进；提升，升级；（商品等的）推广 | 6A |
| propel | [prə'pel] | vt.推进，推动，驱动 | 4A |
| propeller | [prə'pelə] | n.螺旋桨，推进器 | 1B |
| property | ['prɒpəti] | n.特性；属性 | 2A |
| proprietary | [prə'praıətri] | adj.专有的，专利的 | 6B |
| protect | [prə'tekt] | v.保护 | 10A |
| protocol | ['prəutəkɒl] | n.协议 | 1A |
| provenance | ['prɒvənəns] | n.起源，出处 | 10A |
| provider | [prə'vaıdə] | n.供应者，提供者 | 9B |
| provision | [prə'vıʒn] | n.预备，准备；供应 | 3B |
| proxy | ['prɒksi] | n.代理 | 5B |
| purge | [pɜːdʒ] | vt.清除，（使）净化 n.净化 | 10B |
| qualitative | ['kwɒlıtətıv] | adj.定性的 | 3A |
| quantitative | ['kwɒntıtətıv] | adj.定量的；数量（上）的 | 3A |
| quantum-leap | ['kwɒntəm liːp] | n.巨大突破 | 9A |
| quarantine | ['kwɒrəntiːn] | vt.对……进行检疫；隔离 | 10A |
| query | ['kwıəri] | v.查询 n.疑问，询问 | 2A |
| random | ['rændəm] | adj.任意的；随机的 | 7B |
| ransomware | ['rænsəmweə] | n.勒索软件 | 10A |
| readiness | ['redinəs] | n.准备就绪；愿意，乐意 | 10B |
| reapply | [,riːə'plaı] | v.再运用；再申请 | 7B |
| re-assigned | [rı ə'saınd] | adj.重新分配的，重新指定的 | 9B |
| recommendation | [,rekəmen'deıʃn] | n.推荐 | 1B |
| recommender | [,rekə'mendə] | n.推荐引擎；推荐系统 | 2B |
| reconcile | ['rekənsaıl] | vt.使一致 | 5A |
| record | ['rekɔːd] | n.记录；档案 | 6A |
| recover | [rı'kʌvə] | vt.恢复；重新获得；找回 | 5A |

（续）

| 单　词 | 音　标 | 意　义 | 单　元 |
|---|---|---|---|
| recovery | [rɪ'kʌvəri] | n.恢复，复原 | 3B |
| recur | [rɪ'kɜ:] | vi.再发生；复发；重现 | 5A |
| redefine | [ˌri:dɪ'faɪn] | v.重新定义，再定义 | 6B |
| redundancy | [rɪ'dʌndənsi] | n.冗余；多余 | 3B |
| re-engineering | ['ri:ˌendʒɪ'nɪərɪŋ] | n.再设计，重建 v.再造；重建 | 2B |
| refresh | [rɪ'freʃ] | vt.刷新；使恢复 | 5A |
| region | ['ri:dʒən] | n.范围，领域 | 5B |
| regional | ['ri:dʒənl] | adj.地区的，区域的 | 6A |
| registration | [ˌredʒɪ'streɪʃn] | n.登记，注册 | 3A |
| regression | [rɪ'greʃn] | n.回归 | 6A |
| regularly | ['regjələli] | adv.有规律地，按时；整齐地；不断地；定期地 | 3A |
| regulate | ['regjuleɪt] | vt.调节，调整 | 9B |
| regulatory | ['regjələtəri] | adj.管理的；监管的 | 3B |
| reject | [rɪ'dʒekt] | vt.拒绝；抛弃，扔掉 | 5A |
| relation | [rɪ'leɪʃn] | n.关系；联系；关联 | 2A |
| relevant | ['reləvənt] | adj.有关的，相关联的 | 8A |
| reliable | [rɪ'laɪəbl] | adj.可靠的；可信赖的 | 5A |
| remarkable | [rɪ'mɑ:kəbl] | adj.卓越的；显著的 | 7B |
| remediation | [rɪˌmi:dɪ'eɪʃn] | n.补救；纠正 | 10B |
| remodeling | [ri:'mɒdlɪŋ] | v.重构，改造 | 6A |
| remote | [rɪ'məʊt] | adj.远程的，遥远的 | 4A |
| remove | [rɪ'mu:v] | v.消除，移除 | 4A |
| repair | [rɪ'peə] | vt.修理；纠正；恢复 | 1B |
| repeatability | [rɪ'pi:tə'bɪlɪti] | n.可重复性；反复性；再现性 | 2B |
| repercussion | [ˌri:pə'kʌʃn] | n.后果；反响 | 10A |
| replace | [rɪ'pleɪs] | vt.替换；代替 | 1B |
| replicate | ['replɪkeɪt] | vt.复制；重复 | 3B |
| repository | [rɪ'pɒzətri] | n.仓库；储藏室 | 4A |
| requirement | [rɪ'kwaɪəmənt] | n.需求，要求；必要条件 | 6A |
| rescue | ['reskju:] | v.营救，救助 n.营救（行动） | 7A |
| resemblance | [rɪ'zembləns] | n.相似，相似之处 | 2B |
| responsibility | [rɪˌspɒnsə'bɪləti] | n.责任；职责 | 3B |
| responsibly | [rɪ'spɒnsəbli] | adv.负责地；有责任感地 | 10B |
| responsive | [rɪ'spɒnsɪv] | adj.响应的 | 1A |
| restart | [ˌri:'stɑ:t] | v.重新开始 | 5A |

（续）

| 单　词 | 音　标 | 意　义 | 单　元 |
|---|---|---|---|
| retention | [rɪ'tenʃn] | n.保留；保持力 | 10B |
| retrieve | [rɪ'triːv] | vt.检索；重新得到 | 2A |
| reuse | [ˌriːˈjuːz] | vt.再用，重新使用 | 5A |
| rigid | ['rɪdʒɪd] | adj.严格的 | 2A |
| role-based | [rəʊl beɪst] | adj.基于角色的 | 10A |
| rollback | ['rəʊlbæk] | n.回滚 | 5A |
| rotate | [rəʊ'teɪt] | v.（使某物）旋转；使转动 | 1B |
| rough | [rʌf] | adj.粗略的 | 10A |
| routine | [ruː'tiːn] | n.常规；例行程序 adj.例行的；常规的 | 4A |
| row | [rəʊ] | n.行，排 | 2A |
| rule | [ruːl] | n.规则 v.控制 | 2A |
| safeguard | ['seɪfɡɑːd] | n.保护；防护措施；安全设施 vt.保护；防护 | 10B |
| safeguarding | ['seɪfɡɑːdɪŋ] | n.安全措施 | 9B |
| sample | ['sɑːmpl] | n.样本；样品 vt.抽样调查；取样 | 5A |
| sanction | ['sæŋkʃn] | n.制裁，处罚 | 10A |
| satisfy | ['sætɪsfaɪ] | vt.使确信；符合，达到（要求、规定、标准等）vi.使满足或足够 | 6A |
| scalability | [skeɪlə'bɪləti] | n.可扩展性；可升级性；可伸缩性；可量测性 | 10A |
| scalable | ['skeɪləbl] | adj.可伸缩的，可扩展的，可升级的 | 2A |
| scatter | ['skætə] | v.散开，分散 | 4A |
| scene | [siːn] | n.地点，现场；场面 | 7A |
| schedule | ['ʃedjuːl] | n.时刻表，进度表；清单，明细表 vt.排定，安排 | 3A |
| scheduler | ['ʃedjuːlə] | n.调度程序，日程安排程序 | 5B |
| schema | ['skiːmə] | n.概要，纲要；图表 | 2A |
| seamlessly | ['siːmləsli] | adv.无缝地，无空隙地 | 7A |
| search | [sɜːtʃ] | v.&n.搜索；调查 | 2A |
| sector | ['sektə] | n.部门；领域 | 1B |
| seek | [siːk] | vt.寻找，探寻 | 4B |
| segment | ['seɡmənt] | n.部分，段落 v.分段，分割；划分 | 6A |
| select | [sɪ'lekt] | vt.选择 | 4A |
| semantic | [sɪ'mæntɪk] | adj.语义的，语义学的 | 2A |
| sensitive | ['sensətɪv] | adj.敏感的；易受影响的 | 3B |
| sensitivity | [ˌsensə'tɪvəti] | n.敏感；感受性 | 10A |
| sensor | ['sensə] | n.传感器 | 3A |
| sentiment | ['sentɪmənt] | n.感情，情绪；意见，观点 | 6A |

（续）

| 单　　词 | 音　　标 | 意　　义 | 单　元 |
|---|---|---|---|
| separation | [ˌsepəˈreɪʃn] | n.分离，分开；间隔 | 2A |
| sequence | [ˈsiːkwəns] | n.序列；顺序；连续 vt.安排顺序 | 2A |
| serious | [ˈsɪərɪəs] | adj.严重的；重要的；危急的 | 1A |
| server | [ˈsɜːvə] | n.服务器 | 2A |
| shade | [ʃeɪd] | n.阴影；色度 | 8A |
| share | [ʃeə] | vi.分享；共有 | 3A |
| significance | [sɪgˈnɪfɪkəns] | n.重要性；含义 | 8A |
| significant | [sɪgˈnɪfɪkənt] | adj.重要的；显著的 | 7A |
| similarity | [ˌsɪməˈlærəti] | n.类似；相似点 | 6B |
| simplify | [ˈsɪmplɪfaɪ] | vt.简化 | 8B |
| simultaneously | [ˌsɪməlˈteɪnɪəsli] | adv.同时地 | 4A |
| situation | [ˌsɪtʃuˈeɪʃn] | n.情况；形势，处境；位置 | 6A |
| smoothing | [ˈsmuːðɪŋ] | v.（使）光滑，（使）平坦 | 6A |
| sophisticated | [səˈfɪstɪkeɪtɪd] | adj.复杂的；精致的；富有经验的 | 5B |
| spam | [spæm] | n.垃圾邮件 | 5A |
| specifically | [spəˈsɪfɪkli] | adv.特有地，明确地 | 7B |
| spectrum | [ˈspektrəm] | n.光谱，波谱；范围，系列 | 7A |
| sphere | [sfɪə] | n.范围 | 10A |
| spreadsheet | [ˈspredʃiːt] | n.电子制表软件 | 1A |
| stable | [ˈsteɪbl] | adj.稳定的；持久的 | 1A |
| stage | [steɪdʒ] | n.阶段 | 3A |
| stakeholder | [ˈsteɪkhəʊldə] | n.股东；利益相关者 | 5A |
| standard | [ˈstændəd] | n.标准，规格 adj.标准的 | 2B |
| standardize | [ˈstændədaɪz] | vt.使标准化 | 4B |
| state | [steɪt] | n.状态 | 7B |
| statement | [ˈsteɪtmənt] | n.语句；声明 | 4A |
| storage | [ˈstɔːrɪdʒ] | n.存储 | 1A |
| straightforward | [ˌstreɪtˈfɔːwəd] | adj.直截了当的；坦率的；明确的 | 3A |
| strategy | [ˈstrætədʒi] | n.策略，战略 | 1A |
| streamline | [ˈstriːmlaɪn] | vt.使简单化；使现代化 | 10B |
| strenuous | [ˈstrenjuəs] | adj.费力的，用力的 | 9A |
| stretch | [stretʃ] | v.伸展；绵延，延续 | 9A |
| strict | [strɪkt] | adj.精确的；绝对的；严格的 | 3B |
| strike | [straɪk] | v.&n.攻击 | 3B |
| string | [strɪŋ] | n.字符串 | 2A |

（续）

| 单　词 | 音　标 | 意　义 | 单　元 |
|---|---|---|---|
| structure | ['strʌktʃə] | n.结构；构造；体系 vt.构成，排列；安排 | 2A |
| subject | ['sʌbdʒɪkt] | n.主题 v.提出 | 10B |
| subscription | [səb'skrɪpʃn] | n.（报刊等的）订阅费；（俱乐部等的）会员费 | 3A |
| subsequent | ['sʌbsɪkwənt] | adj.后来的，随后的 | 7B |
| subset | ['sʌbset] | n.子集 | 4A |
| sufficient | [sə'fɪʃnt] | adj.足够的，充足的，充分的 | 2A |
| suggest | [sə'dʒest] | v.建议；推荐；表明 | 10A |
| summarize | ['sʌməraɪz] | vt.总结，概述 | 5A |
| supplement | ['sʌplɪmənt] | vt.增强，补充 | 3B |
| surrounding | [sə'raʊndɪŋ] | adj.周围的，附近的 n.周围环境 | 1B |
| surveillance | [sɜː'veɪləns] | n.监视；监督 | 10A |
| survey | ['sɜːveɪ] | n.调查；概述 v.审察，测量；进行民意测验 | 8A |
| suspicious | [sə'spɪʃəs] | adj.可疑的；怀疑的；不信任的 | 10A |
| symbol | ['sɪmbl] | n.符号；标志 | 2B |
| symptom | ['sɪmptəm] | n.症状；征兆 | 1B |
| table | ['teɪbl] | n.表格 v.提交（议案） | 2A |
| tag | [tæg] | n.标签 vt.加标签于 | 2A |
| talent | ['tælənt] | n.人才；天才 | 4A |
| taxonomy | [tæk'sɒnəmi] | n.分类学，分类系统 | 2A |
| tech | [tek] | n.技术 | 10A |
| tenet | ['tenɪt] | n.原则；信条 | 1A |
| terabyte | ['terəbaɪt] | n.太字节，缩写为 TB，1TB=1024GB=$2^{40}$B | 10A |
| tester | ['testə] | n.测试员 | 5A |
| third-party | ['θɜːdpɑːti] | adj.第三方的 | 3B |
| threshold | ['θreʃhəʊld] | n.门槛；阈值；临界值 | 5A |
| tier | [tɪə] | n.级，阶，层；阶层，等级 | 4B |
| tightly | ['taɪtli] | adv.紧紧地，坚固地，牢固地 | 2A |
| time-consuming | ['taɪm kən'sjuːmɪŋ] | adj.费时的，旷日持久的 | 4A |
| timeframe | ['taɪmfreɪm] | n.时间表 | 3A |
| timeline | ['taɪmlaɪn] | n.时间轴，时间表 | 8A |
| toolset | ['tuːlset] | n.工具集，工具包，工具箱；成套工具 | 10A |
| topology | [tə'pɒlədʒi] | n.拓扑 | 5B |
| track | [træk] | vt.跟踪；监看，监测 n.小路；踪迹 | 1B |
| tracking | ['trækɪŋ] | n.跟踪 | 3A |

（续）

| 单　词 | 音　标 | 意　义 | 单　元 |
|---|---|---|---|
| traffic | ['træfɪk] | n.信息流；流量，通信量 | 10A |
| train | [treɪn] | v.训练 | 6A |
| transactional | [træn'zækʃənəl] | adj.交易的，业务的 | 3A |
| transfer | [træns'fɜ:] | v.传输，转移 | 3B |
| transformation | [,trænsfə'meɪʃn] | n.变化；转换；变换 | 4B |
| translatable | [træns'leɪtəbl] | adj.可译的，可转换的 | 5B |
| transparent | [træns'pærənt] | adj.透明的；易懂的 | 10B |
| treatment | ['tri:tmənt] | n.治疗；处理 | 1B |
| tremendous | [trə'mendəs] | adj.极大的，巨大的；可怕的，惊人的 | 9A |
| trend | [trend] | n.趋势；走向 | 6A |
| truncate | [trʌŋ'keɪt] | vt.截断，缩短 | 4A |
| tuple | [tʌpl] | n.元组，数组 | 4A |
| turnoff | ['tɜ:nɒf] | n.避开；岔开 | 7A |
| ubiquitous | [ju:'bɪkwɪtəs] | adj.普遍存在的 | 5B |
| unavailable | [,ʌnə'veɪləbl] | adj.难以获得的；不能利用的 | 10B |
| uncertain | [ʌn'sɜ:tn] | adj.不确定的；不稳定的；不明确的 | 2A |
| unclear | [,ʌn'klɪə] | adj.不清晰的，含糊不清 | 2A |
| uncover | [ʌn'kʌvə] | vi.发现，揭示 | 1A |
| undertake | [,ʌndə'teɪk] | vt.承担，从事；同意，答应 | 2A |
| undoubtedly | [ʌn'daʊtɪdli] | adv.毋庸置疑地，的确地；显然，必定 | 1A |
| uniformity | [ju:nɪ'fɔ:məti] | n.同一性；同样，一样 | 7B |
| unintentional | [,ʌnɪn'tenʃənl] | adj.无意的，无心的 | 3B |
| unlimited | [ʌn'lɪmɪtɪd] | adj.无限的 | 3B |
| unplanned | [,ʌn'plænd] | adj.无计划的，未筹划的 | 9B |
| unsatisfactory | [,ʌn,sætɪs'fæktəri] | adj.不能令人满意的；不符合要求的；不满足的 | 8A |
| update | [,ʌp'deɪt]<br>['ʌpdeɪt] | vt.更新，升级<br>n.更新，升级 | 10A |
| upfront | [,ʌp'frʌnt] | adj.在前面的；预先的 | 4B |
| up-front | ['ʌp 'frʌnt] | adj.预付的 | 5B |
| usable | ['ju:zəbl] | adj.可用的；合用的；便于使用的 | 5A |
| user-friendly | [ju:zə'frendli] | adj.用户友好的，用户容易掌握使用的 | 7B |
| utilize | ['ju:təlaɪz] | vt.利用，使用 | 6B |
| valid | ['vælɪd] | adj.有效的 | 1A |
| validate | ['vælɪdeɪt] | vt.确认；证实 | 10A |

(续)

| 单　词 | 音　标 | 意　义 | 单　元 |
|---|---|---|---|
| validity | [və'lɪdəti] | n.有效性，合法性 | 1A |
| valuable | ['væljuəbl] | adj.有价值的 | 1A |
| variable | ['veəriəbl] | n.变量 adj.变量的；变化的，可变的 | 4A |
| variant | ['veəriənt] | n.变体，变种，变形 | 10A |
| variety | [və'raɪəti] | n.多样化；种类 | 1A |
| vehicle | ['viːəkl] | n.车辆，交通工具 | 1B |
| velocity | [və'lɒsəti] | n.速率，速度；高速，快速 | 1A |
| veracity | [və'ræsəti] | n.真实性 | 1A |
| verbose | [vɜː'bəʊs] | adj.冗长的，啰嗦的 | 7A |
| verbosity | [vɜː'bɒsəti] | n.冗长，赘言 | 7A |
| verification | [ˌverɪfɪ'keɪʃn] | n.验证；证明；证实 | 5A |
| versatility | [ˌvɜːsə'tɪləti] | n.多用途 | 5B |
| virtual | ['vɜːtʃuəl] | adj.虚拟的；实际的 | 3B |
| visualization | [ˌvɪʒuəlaɪ'zeɪʃn] | n.可视化，形象化；虚拟化 | 8A |
| visualize | ['vɪʒuəlaɪz] | vt.使形象化，使可视化 | 5B |
| visually | ['vɪʒuəli] | adv.可视化地；视觉上 | 2B |
| volatility | [ˌvɒlə'tɪləti] | n.易变性 | 1A |
| volume | ['vɒljuːm] | n.容量，大量 | 1A |
| voluminous | [və'luːmɪnəs] | adj.大的；多产的 | 1A |
| vulnerability | [ˌvʌlnərə'bɪləti] | n.弱点；脆弱性 | 10A |
| website | ['websaɪt] | n.网站 | 3A |
| workload | ['wɜːkləʊd] | n.工作量，工作负担 | 5B |
| workspace | ['wɜːkspeɪs] | n.工作区，工作空间 | 5B |

# 附录 C　词组表

| 词　组 | 意　义 | 单　元 |
|---|---|---|
| a bunch of | 一束；一群；一堆 | 6B |
| a lot of | 许多的 | 1B |
| a series of | 一系列；一连串 | 4A |
| a suite of | 一系列；一套 | 5B |
| a variety of | 多种的，种种 | 2A |
| access control | 访问控制 | 10A |
| account number | 账号 | 5A |

（续）

| 词 组 | 意 义 | 单 元 |
|---|---|---|
| adhere to | 遵循；依附；坚持 | 5A |
| after-the-fact method | 事后方法 | 2B |
| agile programming | 敏捷编程 | 2B |
| Apriori algorithm | Apriori 算法 | 6B |
| area chart | 面积图 | 8A |
| artificial intelligence | 人工智能 | 10B |
| association rule | 关联规则 | 6B |
| at rest | 静止，不动 | 10B |
| autonomous database | 自治数据库 | 4A |
| back end | 后端 | 4A |
| bar chart | 条形图 | 8A |
| base on | 基于 | 2A |
| be associated with | 与……相关联 | 4A |
| be available to | 可被……利用或得到 | 9B |
| be aware of | 知道，意识到 | 8A |
| be classified into | 分成，分为 | 9A |
| be compared with | 与……相比较 | 2A |
| be concerned with | 涉及；参与，干预；关心 | 10A |
| be configured to | 被配置为 | 5A |
| be constrained by | 受……约束 | 2A |
| be constructed by | 由……构建 | 6B |
| be converted to | 转换为……，改变为…… | 2A |
| be couple with | 与……一起，连同…… | 2B |
| be declared as | 宣布为 | 7A |
| be described as | 被描述为 | 4A |
| be divided into | 被分为 | 7B |
| be engaged in | 正着手于，正做着，致力于 | 10A |
| be exhibited by | 用……展示 | 6A |
| be factored in | 被考虑进去 | 8A |
| be familiar with | 熟悉的；友好的 | 8A |
| be grouped in | 用……分组，按……分组 | 2B |
| be hosted on | 被托管在……之上 | 4B |

（续）

| 词　组 | 意　义 | 单　元 |
|---|---|---|
| be interested in | 对……感兴趣 | 1B |
| be involved with | 与……有密切关系，涉及…… | 2B |
| be organized as | 被组织成 | 4A |
| be shared with | 与……共享 | 8B |
| be suitable for | 适合……的 | 9A |
| big data | 大数据 | 1A |
| big data cloud | 大数据云 | 9A |
| big data security | 大数据安全 | 10A |
| boosting algorithm | 提升算法 | 6B |
| bottom line | 最终盈利 | 5A |
| brick-and-mortar store | 实体店 | 3A |
| bubble chart | 气泡图 | 8A |
| budding programmer | 新手程序员 | 7A |
| build models | 建立模型，构建模型 | 6A |
| business data | 业务数据，商业数据 | 5A |
| business decision | 业务决策，商业决策 | 5A |
| business intelligence | 商业智能 | 2A |
| call switch | 呼叫交换 | 5A |
| carry out | 执行；进行 | 1B |
| centralized key management | 集中式密钥管理 | 10A |
| cloud data warehouse | 云数据仓库 | 4B |
| cloud database | 云数据库 | 4A |
| cloud provider | 云提供商 | 3B |
| cloud storage | 云存储 | 3B |
| cloud-based database management system | 基于云的数据库管理系统 | 4A |
| clustering algorithm | 聚类算法 | 6B |
| collaborative filtering | 协同过滤 | 7B |
| collective term | 统称 | 10A |
| combine with | 与……结合 | 7B |
| communication protocol | 通信协议 | 5A |
| competitive advantage | 竞争优势 | 1A |
| computation as service | 计算即服务 | 9A |

（续）

| 词　　组 | 意　　义 | 单　元 |
| --- | --- | --- |
| computer program | 计算机程序 | 2A |
| computer system | 计算机系统 | 4A |
| computing model | 计算模型 | 9A |
| conform to | 符合，遵照 | 2A |
| connect the dots | 连点成线 | 10B |
| consist of | 包含；由……组成 | 3A |
| cost efficient | 成本效益 | 9B |
| credit card | 信用卡 | 1A |
| crop up | 犯错误；突然发生；意外地发现 | 10A |
| customer experience | 客户体验 | 5A |
| customer relationship management system | 客户关系管理系统 | 3A |
| cutting edge | 尖端；最前沿；优势 | 5B |
| cyber attack | 网络攻击 | 3B |
| data access control | 数据存取控制，数据访问控制 | 4A |
| data analysis | 数据分析 | 1A |
| data analyst | 数据分析师 | 7B |
| data analytics software | 数据分析软件 | 8B |
| data collection | 数据收集 | 3A |
| data definition | 数据定义 | 4A |
| data flow | 数据流 | 1A |
| data integrity | 数据完整性 | 5A |
| data lake | 数据湖 | 4B |
| data leakage | 数据泄露 | 6A |
| data map | 数据映射 | 5A |
| data mart | 数据集市 | 4B |
| data mining | 数据挖掘 | 1A |
| data model | 数据模型 | 2A |
| data modeling | 数据建模 | 2B |
| data module | 数据模块 | 8B |
| data point | 数据点 | 3A |
| data processing technique | 数据处理技术 | 1A |
| data scientist | 数据科学家 | 7B |

（续）

| 词　　组 | 意　　义 | 单　元 |
|---|---|---|
| data segment | 数据段 | 2B |
| data set | 数据集 | 2B |
| data sharing | 数据共享 | 1A |
| data source | 数据源 | 7B |
| data sovereignty | 数据主权 | 9A |
| data storage | 数据存储 | 1A |
| data stream | 数据流 | 5B |
| data type | 数据类型 | 7B |
| data visualization | 数据可视化 | 1A |
| data visualization tool | 数据可视化工具 | 8B |
| data warehouse | 数据仓库 | 4A |
| database management system | 数据库管理系统 | 9B |
| data-driven company | 数据驱动公司 | 8B |
| data-processing engine | 数据处理引擎 | 7B |
| debit card | 借记卡 | 1B |
| decision stump algorithm | 决策树桩算法 | 6B |
| decision tree | 决策树 | 6B |
| declarative language | 说明性语言；声明性语言 | 4A |
| deep learning | 深度学习 | 7A |
| digital space | 数字空间 | 9B |
| direct move | 直接移动 | 5A |
| dirty data | 脏数据，废数据 | 1A |
| disaster recovery | 灾难恢复 | 9A |
| disk space | 磁盘空间 | 6B |
| distributed computing | 分布式计算 | 7A |
| distributed database | 分布式数据库 | 4A |
| distributed file system | 分布式文件系统 | 9A |
| distributed mode | 分布模式 | 7A |
| distributed virtual file system | 分布式虚拟文件系统 | 9A |
| distribution channel | 分销渠道 | 1A |
| document database | 文档数据库 | 4A |
| drag and drop | 拖放 | 8B |

（续）

| 词　　组 | 意　　义 | 单　元 |
|---|---|---|
| due diligence | 应有的注意，尽职调查 | 10A |
| dynamic masking | 动态屏蔽 | 10B |
| dynamic virtual machine | 动态虚拟机 | 9A |
| e-commerce site | 电子商务网站 | 1B |
| end user | 最终用户，终端用户 | 4A |
| entity-relationship model | 实体关系模型 | 2B |
| evaluation metrics | 评价指标 | 7B |
| external storage device | 外部存储设备 | 3B |
| external threat | 外部威胁 | 10A |
| figure out | 计算出；弄明白；想出 | 9B |
| file-based system | 基于文件的系统 | 2B |
| financial management | 财务管理 | 4B |
| focus on | 致力于；使聚焦于 | 3B |
| follow suit | 跟着做，照着做；如法炮制 | 10B |
| frame diagram | 框架图 | 8A |
| fraud detection | 欺诈检测 | 2B |
| frequent pattern mining | 频繁模式挖掘 | 7B |
| Gantt chart | 甘特图，线条图 | 8A |
| gradient-boosted tree | 梯度引导树 | 7B |
| graph data | 图形数据 | 7B |
| graph data model | 图形数据模型 | 2B |
| graph database | 图形数据库 | 4A |
| graph-based model | 基于图形模式 | 2A |
| graph-parallel computation | 图并行计算 | 7B |
| graph-processing algorithm | 图处理算法 | 7B |
| hard drive | 硬盘驱动器 | 3B |
| heart bit rate | 心率 | 1B |
| heat map | 热图 | 8A |
| hidden pattern | 隐藏模式，隐含模式 | 6A |
| hierarchical data model | 层次数据模型，分级数据模型 | 2B |
| hierarchical structure | 层次结构，分级结构 | 9A |
| high-value customer | 高价值的客户 | 6A |

（续）

| 词　　组 | 意　　义 | 单　元 |
|---|---|---|
| historical data | 历史数据 | 4B |
| hybrid cloud | 混合云 | 3B |
| in a short period | 短期内 | 5A |
| in anticipation of | 期待着……；预计到…… | 3B |
| in order to | 为了…… | 5A |
| in the form of | 用……的形式 | 6B |
| in transit | 传输中的；在途中 | 10A |
| in various cases | 在各种情况下 | 1B |
| information processing | 信息处理 | 1A |
| integrate system | 整合系统，集成系统 | 5A |
| integrate with | 与……结合，与……集成 | 5B |
| intend to | 打算，期望 | 4A |
| internal attack | 内部攻击 | 9B |
| Internet of Things | 物联网 | 1A |
| inventory management | 库存管理 | 1A |
| iterative development | 迭代开发 | 7A |
| keep up with | 跟上 | 4A |
| large-scale distributed computing | 大规模分布式计算 | 9A |
| lazy learning algorithm | 惰性学习算法，消极学习算法 | 6B |
| lead to | 导致 | 5A |
| learning curve | 学习曲线 | 7A |
| limit to | （把……）限制在……，局限于…… | 10A |
| line chart | 线形图，折线图 | 8A |
| linear algebra | 线性代数 | 7B |
| link analysis algorithm | 链接分析算法，链路分析算法 | 6B |
| litigation cost | 诉讼费 | 10A |
| lock down | 锁定，保持 | 10B |
| logical data model | 逻辑数据模型 | 2B |
| logical partition | 逻辑分区 | 7B |
| machine learning | 机器学习 | 4A |
| maintenance mode | 维护模式 | 7B |
| make up | 组成；补足 | 3A |

（续）

| 词 组 | 意 义 | 单 元 |
| --- | --- | --- |
| make up for | 弥补 | 7A |
| market share | 市场份额 | 10B |
| markup language | 标记语言，标识语言 | 2A |
| master/slave architecture | 主/从结构 | 7B |
| mobile device | 移动设备 | 1A |
| monitoring device | 监控设备 | 1A |
| monthly subscription service | 每月订购服务 | 9B |
| multidimensional data visualization | 多维数据可视化 | 8B |
| multimodel database | 多模式数据库 | 4A |
| multiple user | 多用户 | 4A |
| mutual relationship | 相互关系 | 6B |
| native code | 本机代码，本地代码 | 5B |
| natural disaster | 自然灾害 | 9B |
| natural language processing | 自然语言处理 | 5B |
| negatively affect | 负面影响，消极影响 | 10A |
| network configuration | 网络配置 | 9A |
| network data model | 网络数据模型 | 2B |
| neural network | 神经网络 | 6A |
| nonrelational database | 非关系数据库 | 4A |
| non-sensitive task | 非敏感任务 | 9B |
| object-oriented database | 面向对象数据库 | 4A |
| object-oriented modeling | 面向对象建模 | 2B |
| object-oriented programming | 面向对象程序设计 | 4A |
| on-demand self-service | 按需自助服务 | 9B |
| online storage | 网络存储 | 9B |
| open source | 开源 | 4A |
| open source community | 开源社区 | 7B |
| operating system | 操作系统 | 4B |
| opex-based delivery model | 基于运营的交付模型 | 9B |
| originate from | 来自……，源于…… | 10A |
| out-of-the-box solution | 开箱即用的解决方案 | 7B |
| pain point | 痛点 | 8A |

（续）

| 词　　组 | 意　　义 | 单　元 |
|---|---|---|
| parallel data processing | 并行数据处理 | 7B |
| parallel processing | 并行处理 | 7A |
| parent class | 父类 | 2B |
| pay-as-you-go model | 随用随付模式 | 9B |
| pertain to | 属于；关于；适合 | 10A |
| physical data model | 物理数据模型 | 2B |
| physical infrastructure | 物理基础设施 | 9A |
| pie chart | 饼图 | 8A |
| pile up | 堆积，积聚，成堆 | 1A |
| platform agnostic | 平台无关的 | 10B |
| privacy regulation | 隐私法规 | 10B |
| private cloud | 私有云，专用云 | 3B |
| private network | 私有网络，专用网络 | 3B |
| processing unit | 处理单元 | 6A |
| programming language | 程序设计语言，编程语言 | 4A |
| proxy server | 代理服务器 | 5B |
| public cloud | 公共云 | 3B |
| random forest | 随机森林 | 7B |
| rapid development methodology | 快速开发方法 | 2B |
| real time | 实时 | 1A |
| relational data model | 关系数据模型 | 2B |
| relational database | 关系数据库，关系型数据库 | 2A |
| reside in | 住在……，存在于…… | 2A |
| respond to | 对……做出反应 | 10B |
| response time | 响应时间 | 5A |
| reverse engineering | 逆向工程，反向工程 | 2B |
| risk exposure | 风险敞口 | 10B |
| sample data | 样本数据 | 5A |
| scale up | 按比例增加，按比例提高 | 3B |
| scatter plot | 散点图，散布图 | 8A |
| scripting language | 脚本语言 | 7A |
| search engine | 搜寻引擎 | 6B |

（续）

| 词　组 | 意　义 | 单　元 |
|---|---|---|
| secret sauce | 秘诀 | 10B |
| security service level agreement | 安全服务级别协议 | 10A |
| self-driving database | 自动驾驶数据库 | 4A |
| self-service analytics tool | 自助分析工具 | 8B |
| semi-structured data | 半结构化数据 | 2A |
| sensitive information | 敏感信息 | 10A |
| serve as | 充当，担任 | 2B |
| service provider | 服务提供商 | 4A |
| set up | 建立，设立；安排 | 3A |
| sign up for | 报名；注册 | 3A |
| smart assistant | 智能助理 | 8B |
| smart electric meter | 智能电表 | 1B |
| smart traffic system | 智能交通系统 | 1B |
| snowflake schema | 雪花模式 | 2B |
| social media | 社交媒体 | 1A |
| social network | 社交网络 | 10B |
| source code | 源代码 | 4A |
| spending habit | 消费习惯 | 1B |
| staging area | 暂存区域，临时区域 | 5A |
| stand out | 突出；出色；更为重要；引人注目；显眼 | 6A |
| star schema | 星型模式 | 2B |
| statistical computing | 统计计算 | 7A |
| storage cost | 存储成本，存储花费 | 2A |
| storage repository | 存储库 | 9A |
| streaming application | 流媒体应用 | 7A |
| streaming data | 流数据 | 7B |
| strong learner | 强学习器 | 6B |
| structured data | 结构化数据 | 2A |
| supply chain | 供应链 | 1A |
| take ... for granted | 认为……理所当然，想当然；轻信 | 10A |
| take place | 发生 | 9A |
| target audience | 目标受众 | 3A |

（续）

| 词　　组 | 意　　义 | 单　元 |
|---|---|---|
| target database | 目标数据库 | 4B |
| technical skill | 技术技能；专门技能 | 5A |
| text analysis | 文本分析 | 2B |
| text file | 文本文件 | 5A |
| text report | 文本报告 | 8B |
| third party | 第三方 | 3A |
| third-party data center | 第三方数据中心 | 9B |
| third-party service | 第三方服务 | 9A |
| tier-conscious strategy | 层级意识策略 | 10A |
| town planner | 城市规划者 | 6A |
| trade off | 权衡，交易 | 5A |
| transformation engine | 转换引擎 | 4B |
| unlabeled data | 未标记的数据 | 6B |
| unstructured data | 非结构化数据 | 2A |
| use case | 用例 | 5B |
| user access control | 用户访问控制 | 10A |
| utility computing | 效用计算 | 9B |
| virtual compute | 虚拟计算 | 9A |
| virtual machine | 虚拟机 | 9A |
| virtual personal assistant | 虚拟个人助理 | 1B |
| virtualized infrastructure | 虚拟化基础设施 | 9A |
| visual data model | 可视数据模型 | 8B |
| visualization platform | 可视化平台 | 5B |
| weak learner | 弱学习器 | 6B |
| web page | 网页 | 2A |
| web services application | 网络服务应用程序 | 8B |
| well-thought out | 深思熟虑 | 1A |
| workload portability | 工作负载可移植性 | 9A |
| worst-case scenario | 最坏的情况 | 9B |
| zipped file | 压缩文件 | 2A |

# 附录 D　缩写表

| 词　　组 | 意　　义 | 单　元 |
|---|---|---|
| 2D (2-Dimensional) | 二维 | 5B |
| 3D (3-Dimensional) | 三维 | 5B |
| 4GL (Fourth Generation Language) | 第四代语言 | 4A |
| AI (Artificial Intelligence) | 人工智能 | 4B |
| API (Application Programming Interface) | 应用程序编程接口 | 7B |
| BDF (Big Data Fabric) | 大数据结构 | 9A |
| BI (Business Intelligence) | 商务智能 | 5A |
| BLOB (Binary Large OBject) | 二进制长对象 | 2A |
| CART (Classification and Regression Trees) | 分类与回归树 | 6B |
| CAS (Content Addressable Storage) | 内容可寻址存储 | 2A |
| CDBMS (Columnar DataBase Management System) | 列式数据库管理系统 | 4A |
| CI (Cloud Infrastructure) | 云基础设施 | 9A |
| CMS (Content Management System) | 内容管理系统 | 10A |
| CODASYL (Conference on Data Systems Languages) | 数据系统语言会议 | 2B |
| CPU (Central Processing Unit) | 中央处理器 | 4B |
| CRAN (Comprehensive R Archive Network) | 综合 R 档案网络 | 7A |
| CRM (Customer Relationship Management) | 客户关系管理 | 1A |
| CSV(Comma-Separated Values) | 逗号分隔值，也称为字符分隔值 | 8B |
| DAG (Directed Acyclic Graph) | 有向无环图 | 7B |
| DBaaS (DataBase as a Service) | 数据库即服务 | 4A |
| DBMS (DataBase Management System) | 数据库管理系统 | 2A |
| DDL (Data Definition Language ) | 数据定义语言 | 4A |
| DDoS (Distributed Denial of Service) | 分布式拒绝服务 | 10A |
| DML (Data Manipulation Language) | 数据操作语言 | 4A |
| DMP (Data Management Platform) | 数据管理平台 | 3A |
| DQL (Data Query Language) | 数据查询语言 | 4A |
| ELT (Extract, Load, Transform) | 提取、加载、转换 | 4B |
| EM (Expectation-Maximization) | 期望最大化 | 6B |
| ER (Entity-Relationship) | 实体关系 | 2B |
| ETL (Extract, Transform, Load) | 提取、转换、加载 | 4B |
| EUGDPR (European Union's General Data Privacy Regulation) | 欧盟的通用数据隐私法规 | 10B |

| 词　　组 | 意　　义 | 单　元 |
|---|---|---|
| FAQ (Frequently Asked Questions) | 常见问题，经常问到的问题 | 8B |
| GDPR (General Data Protection Regulation) | 通用数据保护条例 | 1A |
| GIF (Graphics Interchange Format) | 图像互换格式 | 2A |
| GIS (Geographic Information System) | 地理信息系统 | 2B |
| GPS (Global Position System) | 全球定位系统 | 2A |
| GUI (Graphical User Interface) | 图形用户界面，图形用户接口 | 5B |
| HDFS (Hadoop Distributed File System) | Hadoop 分布式文件系统 | 5B |
| HTTP (Hyper Text Transfer Protocol) | 超文本传输协议 | 5B |
| IaaS (Infrastructure as a Service) | 基础设施即服务 | 9B |
| IDS (Intrusion Detection System) | 入侵检测系统 | 10A |
| IMDBMS (In-Memory DataBase Management System) | 内存数据库管理系统 | 4A |
| IMS (Information Management System) | 信息管理系统 | 2B |
| IoT (Internet of Things) | 物联网 | 10B |
| IPS (Intrusion Prevention System) | 入侵预防系统 | 10A |
| IT (Information Technology) | 信息技术 | 1A |
| JPEG（Joint Photographic Experts Group） | 联合图像专家组 | 2A |
| JSON (JavaScript Object Notation) | JS 对象表示法 | 4A |
| JVM (Java Virtual Machine) | Java 虚拟机 | 7A |
| KDD (Knowledge Discovery in Data) | 数据知识发现 | 6A |
| KNN (K-Nearest Neighbor) | K 最近邻算法 | 6B |
| MLlib (Machine Learning library) | 机器学习库 | 7B |
| NIST(National Institute of Standards and Technology) | 国家标准技术研究院 | 9B |
| OEM (Object Exchange Model) | 对象交换模型 | 2A |
| OLTP (On-Line Transaction Processing) | 联机事务处理 | 2A |
| PaaS (Platform as a Service) | 平台即服务 | 9A |
| PII (Personally Identifiable Information) | 个人可识别的信息 | 10B |
| PNG (Portable Network Graphics) | 便携式网络图形 | 2A |
| POS (Point of Sale) | 销售终端 | 5A |
| POSIX (Portable Operating System Interface of UNIX) | UNIX 可移植操作系统接口 | 5B |
| RDBMS (Relational DataBase Management System) | 关系数据库管理系统 | 2A |
| RDD (Resilient Distributed Dataset ) | 弹性分布式数据集 | 7B |
| REST (REpresentational State Transfer) | 表现层状态转移，表述性状态传递 | 9A |

| 词　　组 | 意　　义 | 单　　元 |
|---|---|---|
| RFID (Radio Frequency IDentification) | 射频识别 | 2A |
| ROI (Return on Investment) | 投资利润，投资回报率 | 1A |
| RPC (Remote Procedure Call) | 远程过程调用 | 5B |
| SaaS (Software as a Service) | 软件即服务 | 9B |
| SAN (Storage Area Network) | 存储区域网络 | 3B |
| SLA (Service Level Agreement) | 服务等级协议，服务级别协议 | 5B |
| SQL (Structured Query Language) | 结构化查询语言 | 2A |
| SVM (Support Vector Machine) | 支持向量机 | 6B |
| TCP/IP (Transmission Control Protocol/Internet Protocol) | 传输控制协议/网际协议 | 2A |
| XML (eXtensible Markup Language) | 可扩展标记语言 | 2A |
| XOLAP (eXtended On-Line Analytic Processing) | 扩展联机分析处理 | 2A |
| XSS (Cross Site Scripting) | 跨站脚本攻击，为了与 CSS（Cascading Style Sheets）区别，改了缩写 | 10A |